TECHNOLOGY HUMAN VALUES AND LEISURE

TECHNOLOGY HUMAN VALUES AND LEISURE

Edited by Max Kaplan
and Phillip Bosserman

Abingdon Press
Nashville / New York

TECHNOLOGY, HUMAN VALUES, AND LEISURE

Copyright © 1971 by Abingdon Press

ISBN 0-687-41190-4 (cloth)
ISBN 0-687-41191-2 (paper)

Library of Congress Catalog Card Number: 76-160798

The lines from "If" by Rudyard Kipling are taken from *Reward and Fairies,* and are used by permission of Mrs. George Bambridge and Macmillan & Co.

The lines from "For John F. Kennedy His Inauguration," by Robert Frost, are from *The Poetry of Robert Frost*, edited by Edward Connery Lathem. Copyright © 1961, 1962 by Robert Frost. Copyright © 1969 by Holt, Rinehart and Winston, Inc. Reprinted by permission of Holt, Rinehart and Winston, Inc.

MANUFACTURED BY THE PARTHENON PRESS AT
NASHVILLE, TENNESSEE, UNITED STATES OF AMERICA

PREFACE

The major point made in these pages is that the concern with leisure in the present world is not with one segment of the population—the aged; nor with one fragment of the problem —how many people watch TV; nor with one problem issue—what should one do in his leisure time; nor with one culture—the United States of America; nor with one historical period—our own; nor even with one set of disciplines—the social sciences. This book has grown out of a conference on technology, human values, and leisure held on the campus of the University of South Florida. The breadth of backgrounds of the conference participants demonstrates that the study of leisure covers all ages, a large range of issues far beyond description counts of activities, a conceptualization of public policy, a world problem which has caught the attention of many nations; it moves from the past and present directly into a philosophy for the post-industrial society; it is altogether an interdisciplinary endeavor. The backgrounds of the distinguished participants dramatize the last point: two specialists in leisure research, two philosophers of science, a historian, a public servant, a labor leader, a musician, an economist, two educators, and a humanistic sociologist.

Economist Robert Theobald opens the volume with a way of thinking about the future. He argues that even such terms as "technology," "human values," and "leisure" are parts of an industrial-era language that is obsolete. The paper is therefore a spelling-out of the new language.

Yet a consideration of the future requires more than new kinds of thought. The point is made by Emmanuel Mesthene, the Harvard philosopher of science, that new values are required. What one historian, the late Robert Hilliard, heard in Mesthene's paper was this: "Aristotle and St. Augustine excoriate Isaac Newton, Emile Descartes, Henry Ford, Marshall McLuhan, and every scientist who runs every problem in a computer because he has a computer; every artist who takes up airbrush technique and thinks that thereby he is doing something important; every minister and divine who abandons his more important function for social engineering. Professor Mesthene has charged us with the crime of Esau: we have sold our birthright for a mess of porridge, by forgetting that we are something more than the mere slaves of the machines."

The scientist-author Harrison Brown, of Cal Tech, then outlines in broad strokes how man came to his present "high energy civilization." Are there any limits to the capacity of this civilization to produce all that man needs? None, according to Brown; his small scenario of the possible future may remind the reader of Vonnegut's *Player Piano,* depicting the elites and the nonworkers in tomorrow's city of Ilium. Turning to the present, Brown summarizes the differences between the worlds of the rich and the poor. The disproportionate resources we are putting into arms production may limit our days, and it could be, he observes, that there "won't be any leisure because there won't be any world to have leisure in."

From these broad perspectives by an economist, a philosopher, and a scientist, the emphasis in the next section shifts to implications more specifically related to leisure. It is not surprising that Arthur Schlesinger begins with an eloquent analysis of our society and the commitment of the government to its constructive values; but then, relying upon his close contact with one of President Kennedy's contributions, he turns at length to the importance of the arts as the bridge to the issues of leisure. He notes that a national cultural policy conforms to the early traditions of the nation, but acquires a "new point" in the age of electronic revolution.

The arts and other prototypes of leisure are relatively new concerns of the American labor movement as well as of government; one of the AFL-CIO's important leaders in this forward direction is Leo Perlis, who reviews the labor movement's record in rendering public service. This new direction, he argues, is a natural outgrowth of the new automation factor in work, freeing the worker for a more effective use of his new free time.

However, all this may be getting ahead of ourselves, argues Dorothy Maynor. The struggle for minimum conditions of life are not yet won for a large segment of our people. For the slum resident, she suggests, "the whole concept of leisure is not a positive thing at all." Yet being the great musician that she is, Miss Maynor herself, as the creator of an important arts program in Harlem, is committed to raising the quality of life among her people; this program is described in her present paper.

No knowledgeable American can be unaware of Robert Hutchins' long dialogue with the institutions of higher learning. As always, he too pleads for a society of quality, a "learning society" that is now made possible by a science and a technology that are here to stay. The summary of his Center for the Study of Democratic Institutions after ten years of activity in Southern California may itself serve as a prototype for one form of creative leisure, serving the purpose "of independent thought and criticism, to learn how to gain comprehension through dialogue, to clarify the basic issues—and some burning ones—and to widen the circles of discussion about them."

It is a natural step from Hutchins' major comment on education to a more specific discussion on youth. Phillip Bosserman's analysis, prepared for this volume after the symposium itself was held, grows out of a unique aspect of the Tampa meeting. Early in the planning of the meeting, several leading students of the University were invited to spend two afternoons in public dialogue with several of the guests. Oddly, we thought, but in fact, these students asked to have with them on the platform several faculty to be chosen by them.[1] The discussions were a highpoint of the symposium, for at this time the student movement across the country was well along. Rather than reproduce or summarize these long meetings, Professor Bosserman has absorbed the spirit of responses by alert students of today to the leisure question which was evident in these discussions; he has, however, clothed his impressions of those discussions with his own analysis of youth.

A slightly different version of the material in this chapter was presented to the White House Conference on Youth, Task Force on Work and Economy, Spring, 1971. The "working paper" format has been retained.

The volume next turns to the work of two leading scholars carrying on research in the leisure field, from eastern and western Europe. Alexander Szalai's organizational genius as well as his impeccable standards of research carried to a conclusion the most successful perhaps of all international team efforts among social scientists. This 12-nation time-budget will be reported in two volumes being readied for publication; since some selected conclusions of the study will appear in full detail, permission was obtained from Dr. Szalai to reproduce here his introductory chapter describing the concepts and methods of such research.

Joffre Dumazedier, the most energetic and productive of all scholars in leisure, also has headed several international groups, including a current nine-nation

[1] "Confronting" Szalai and Brown were Scott Barnett and Walt Terrie, students, and Prof. Graham Solomon of Chemistry; confronting Schlesinger, Theobald, and Perlis were Robert Van Hook and Michael Elworth, students, and Prof. R. Warner (American Idea) and Prof. David Leonard (History).

team of which the Tampa center is a part. Details of his own field research, as in Annecy, can be read in his volume *Toward a Society of Leisure*. For our purposes he has summarized not only some major research in the leisure field but his own positions in a study of the post-industrial society.

Finally, Professor Emeritus Charles Obermeyer, who heard the full symposium, was invited to respond in any way he saw fit to all the papers.

The reader should know that no titles were assigned to the symposium participants. Each had been asked to consider, from his own wide experience, three questions:

1. What is the technology of the post-industrial era, and what are its implications for fundamental changes in our political, economic, and social institutions and patterns?
2. What do these trends mean for the mind, the moral values, the "human condition" of man?
3. What are the complex relationships between leisure and technological growth, and how can these relationships be utilized to promote creative social change?

MAX KAPLAN

CONTENTS

IMPLICATIONS OF LEISURE

RESEARCH INTO LEISURE

FINAL OBSERVATIONS

APPENDIX A: THE ALLOCATION OF LEISURE

APPENDIX B: FREE CITY FESTIVAL POSTER

THE CONTRIBUTORS

PHILLIP BOSSERMAN. Research director, Center for Studies of Leisure, professor of sociology, University of South Florida; studies of education in Liberia; editor, *African Images of Man;* author of *African Mirrors, American Images* and *Dialectical Sociology: A New Approach to the Study of Social Reality.*

HARRISON BROWN. Professor of science and government, California Institute of Technology; author of *The Challenge of Man's Future, The Next Hundred Years* and *A World Without War.*

JOFFRE DUMAZEDIER. Director and chairman, UNESCO Commission on Leisure and Culture; member, Centre d'Etudes Sociologiques; author of *Toward a Society of Leisure.*

15

ROBERT M. HUTCHINS. President, Center for the Study of Democratic Institutions; formerly president of the University of Chicago; associate director of the Ford Foundation; author of many books on higher education.

MAX KAPLAN. Director, Center for Studies of Leisure, University of South Florida; author, lecturer, consultant, sociologist, musician; author of *Leisure in America* and *Foundations and Frontiers of Music Education.*

DOROTHY MAYNOR. Director, The Harlem School of the Arts, New York City; for many years one of the world's great concert sopranos.

EMMANUEL G. MESTHENE. Director, Program on Technology and Society, Harvard University; faculty, Graduate School of Business Administration; research associate, John F. Kennedy School of Government.

CHARLES OBERMEYER. Psychologist and philosopher; professor emeritus, University of South Florida; author of *Body, Soul and Society.*

LEO PERLIS. National Director, AFL-CIO Department of Community Services; a founder and former secretary of CARE; former organizer for Textile Workers Union of America and other unions.

ARTHUR SCHLESINGER, JR. Albert Schweitzer Professor of the Humanities, the City University of New York; Pulitzer Prize historian; special assistant to President Kennedy and President Johnson.

ALEXANDER SZALAI. Deputy Director of Research, United Nations Institute for Training and Research (UNITAR); director of the Multi-National Comparative Time Budget Research Project of the European Coordination Centre for Research and Documentation in Social Sciences.

ROBERT THEOBALD. Noted British-American economist; lecturer; author of *The Challenge of Abundance, The Guaranteed Income,* and *An Alternative Future for America.*

ISSUES

MAX KAPLAN

THE RELEVANCY OF LEISURE

What is relevancy? What are priorities? To the blacks of America, their struggle for jobs, schooling, housing, health, and dignity has no equal claim on our nation—and who is to deny that stand?

To the youth of the nation, in league with students in France, Mexico, Japan, England, and elsewhere, the Indochina war is only one evidence of decadence, middle-class shallowness, and alienation of the academic community from off-campus realities—and who can deny some validity in that stand?

But others as well can stake out claims to their own priorities: the young girl who needs love, the politician who needs votes, the mathematician who needs an equation, the clerk who needs nothing as much as a more sympathetic employer. The priorities of the world move back and forth between the most individualistic, personal, and private, and the most universal, collective, and public.

A large priority at present in the United States is the removal of hunger. It is necessary at the outset in a book about leisure to set that fact straight. On the eve of the White House Conference on Food, Nutrition, and Health in December, 1969, four special committees urged that President Nixon declare a national emergency on this matter, arguing that all other problems of nutrition fade into significance beside the fact that 25 million Americans or more do not have enough to eat. It will bode no good for those concerned with leisure to ignore the kinds of data found in recent volumes by Lundberg, Harrington, Miller, and others.

Yet this is a land and a world of relative as well as absolute deprivation. Compared to generations of the past, large masses of our people in the United States—perhaps as many as 175 million—for whom there is no emergency of hunger, are rather overfed, overclothed, oversupplied with the comforts and the gadgets. In all, our abundance is such that our spending for recreation alone is expected to reach over $46,000 million by 1985. Distribution and intelligent consumption are our problem, not production, for by that same year we could (according to economist Juanita Kreps of Duke) work half a day, half a year, or half a lifetime without lowering the levels of life.

There is need to be deeply concerned with the immediate problems, yet there is also the need to look ahead to the meaning of our whole society. Perhaps, with the view in mind from the Sea of Tranquillity on the moon, we need not only to look ahead but to look around. The panorama from the armchair projections of a few years back is changing dramatically. It is now doubly necessary to look around at man who, in the view of Jacques Ellul,

lives in a universe of concrete, glass, steel, and asphalt, and no longer in a universe of earth, trees, and water. Incessantly he uses tools made of artificial materials and no longer those made of natural materials such as wood and iron. He moves at a speed which is no longer natural speed, whether it be his own or that of a horse. He lives in a densely populated environment, belongs to a variety of groups, and no longer dwells in small autonomous and multifunctional communities. His body no longer defends itself against disease by spontaneous reaction, but by the use of external therapeutics, chemical or surgical. He does not live according to the biological rhythm of the body or of the natural environment, but according to the mechanical rhythm based on a division of time by the clock. At his discretion, he can turn night into day, and live by night. He can be cold in summer and warm in winter.[1]

These, says the French sociologist-historian, are roots of difficulties rather than causes for celebration, especially as they lead now into a second and third degree of technology—techniques now acting upon other techniques, such as the developments in coordinating, classifying, and documenting knowledge.

Where, then, is reality? In *man*, as Society Writ Small, or in *society*, as Man Writ Large? Does a philosophy for our time have to choose between the microcosmic and the macrocosmic? Is there no middle range in cultural philosophy comparable to social psychology

[1] "Technique, Institutions, and Awareness," *American Behavioral Scientist*, XI, July-August, 1968, p. 38.

within the constellation of science, which can deal with
man and with men at the same time?

Leisure is such a middle range. At one time it deals
with the momentary fragment of life: someone sits
quietly and knits, another reads, a third listens to a
symphony or records, a fourth drives his family into
the countryside to celebrate a New England fall day.
At the same time the subject of leisure is as magnificent in its dimensions as the institutions and values of
the culture or the milieu: the matter of loneliness, the
nature of audiences, and the meaning of mass media,
the roots of family life, the perception of time, the uses
of technology, the psychological-economic dimensions
of the guaranteed annual income, the symbols of social
class, the motivations and satisfactions in work. What
is momentary and private is not unimportant, and what
is of grand design to social inquiry is not impractical.
We must, in this field, deal with many realities and
many disguises: the pragmatic and the symbolic, the
committed and the parasitic, the spurious and the au-
thentic.

Vietnam will eventually be resolved, just as the shape
of the negotiating table was; and, hopefully, the Negro
will find his full justice and opportunity. Other issues
will by then have arisen to demand immediate commitments. But even more pressingly, the issues raised by
Fromm and Ellul (and Eric Hoffer, Lewis Mumford,
Don Michael, and others) will push their way into the
forefront of political, social, economic, educational,
philosophical, and even theological issues. Where, indeed, do we head in a world of which David Riesman
asks, *Abundance for What?* That we face a problem
rather than a collective celebration is an enormous ad-

mission, and it is a problem far greater than crime or the decay of the American city. The latter are only dislocations, wrong priorities in national budgets, or (in the international area) what Senator Fulbright calls an arrogance of power. But to come to leisure and to confront it as a problem means that man with machines has made extensions of his body and now finds no more reality in his body. It means that he possesses the imagination to invent a heaven suitable for after-death and create a major institution to pray for deliverance to his vision, but given the chance to bring heaven to his more immediate grasp, he is locked in by his fantasy.

On the first page of an important recent book, John McHale writes:

It is significant that, at this critical juncture in human affairs, man now turns to "ecology" as a guide towards rethinking his overall relationship to his environment—rather than to the more traditional political and economic viewpoints which have guided and measured his large-scale actions before. . . . From the roots of "house-knowledge," we can assume a definition of applied human ecology as "planetary housekeeping." Such an ecological reorientation may, paradoxically, be as disruptive of man's traditional attitudes and values as his physical actions have been in environment.[2]

Not only do political and economic thought need to be reexamined. The "housekeeping" to which McHale alludes goes beyond the physical, the relation of man to his resources in the air, on the land, or in the sea. The same technical and social forces that change the nature of man's habitat are also digging deeply into

[2] John McHale, *The Ecological Context* (New York: George Braziller, 1970), p. 1.

what he does in it and how he thinks about what he does; but mostly *the new physical forces documented by McHale and others provide man with new potentials for access to the goals he desires.* The term leisure is a shorthand way of viewing the new potentials. The new leisure—new in its availability for the masses and the new material base below it—is a convergence of many factors, such as new transportation and communications, cybernation, urbanization, and, on the other hand, of the new questions being raised in the realms of power, justice, purposes, and the meaning of technology. Later in this volume, Phillip Bosserman speaks of three discretionary features which intersect to create a "new world perspective, and new consciousness which is the hallmark of the leisured society." He goes on to say:

Discretionary income can be spent for those items not considered basic needs; discretionary time means time from work; discretionary social behavior opens a myriad of lifestyles to anyone regardless of family, education, age, wealth, ethnic background, and location. This is a new type of society.[3]

As the physical, ecological aspects should have been foreseen, and were by such men as Ellul and Mumford, so the social revolution has been discernible to those who cared to look at basic changes (Robert Theobald, David Riesman are examples). Yet for the masses, the transformation from a society of scarcity to one of abundance has been overtaking them item by item, or, as one writer calls it, a "hair-by-hair" transformation

[3] In the chapter entitled "Implications for Youth," p. 147.

of man's face which ultimately changes his whole appearance.[4]

The masses cannot be asked to become social commentators, for they are busy living; Eric Hoffer raises questions about the roles and, therefore the perspectives of intellectuals; and social scientists who teach us about the cultural lag often fail to see their own lag in catching up with events. Sincere men of all walks still question the seriousness of the changes. Instead of looking at the rate of computerization they count the number of hours that the worker still remains on the job, or moonlights, or expends his material needs. Instead of doubting the precision of the developing discipline of "futurology" they should consider the dramatic compression of technical and social trends, as in the knowledge industry.

In the physical field, the "environmental" problem suddenly becomes political hay; in the social field, the "leisure" problem still remains under the surface, with its proponents in about the same visibility as the environmentalists were two decades ago. Yet both issues interweave, and the transformations which already display themselves and which will emerge in the next few decades can shake to the roots our institutions of family or state, and can create an economic move toward cybernetic productivity that will require a drastic departure from the work ethic, old attitudes toward "welfare" or nonwork maintenance; nothing less than a new philosophy will suffice to provide an ideological scaffolding, and the youth of our present are moving us toward that futuristic construction.

[4] Dr. Lawrence Payne, in a private volume, *Not by Affluence Alone*, written for the International Publications Corp. of London, 1969.

The future impels us to look at our credentials of the past. As Eric Hoffer writes in *The Passionate State of Mind,*

> To enter the realm of the future is like entering a foreign country: one must have a passport, and one must be able to provide a detailed record of one's past. Thus a nation's preoccupation with history is not infrequently an effort to obtain a passport for the future. Often it is a forged passport.[5]

What Hoffer says of a nation applies to an individual as well. Before me is the visa form sent me in preparation for a trip I took to Czechoslovakia recently. On it one indicates where he plans to go; why he is going, how he is going, the place of his birth, and who he is—his occupation. These are not unreasonable questions.

The familiar ticket to life—work—seems in danger of having outlived its acceptability. Going into the land of "cyberculture" and the unfamiliar social and technological terrain of the post-industrial society, the spaceship Earth confronts a collective visa form which must be completed for the journey into the future: where, why, how, who, place of origin? A collective concern with leisure as a new source of meaning and value will color each of the visa entries. We had better not forge the visa; the trip is too important.

[5] (New York: Harper & Row, 1955), p. 48.

PERSPECTIVES BEYOND LEISURE

ROBERT THEOBALD

THINKING ABOUT THE FUTURE

I'll start with three quotations.[1] I'll quote first from a speech I made just a year ago. I argued: "We have something like six to nine months to make visible the beginning of a change from a society of coercive authority to a society of shared power. If trends continue to develop as they are presently developing we will move into a fascist police state in this country. Let me make it clear that I am not arguing that anyone wants a fascist police state. There are few evil men around: our problem is a lack of imagination rather than a problem of evil. We are being forced toward a fascist police state by events,

[1] The informal style of Dr. Theobald's paper has been preserved for publication.

and we will continue to be forced by events until we change our attitudes. The fact that the development of such a police state will be unwilled will not make it any less real."

Now let me quote from a newspaper article. "Attorney General John M. Mitchell called last night for an end to minority tyranny on the nation's campuses and demanded a crackdown on student militants by college officials, law enforcement agencies, and courts. 'If arrests must be made, then arrests there should be,' said Mitchell. 'If violators must be prosecuted then prosecutions there should be. This administration has tried to be patient,' he said, 'in the hopes that students, faculty and local officials working together would put an end to this chaos. But the time has come for an end to patience. The time has come to demand in the strongest possible terms that University officials, local law enforcement agencies, and local courts apply the law.' "[2]

There is one major problem with that statement. As we all know, the law is a very flexible instrument. The law is strongly influenced by what people feel the law ought to be. This pattern is most obvious when a particularly unpleasant murder case has to be moved out of a town. But similar problems arise whenever passions run high. If Attorney General Mitchell continues to make these statements, he must inevitably damage respect for the law. He argued, for example, that it is no admission of defeat to use reasonable physical force to eliminate physical force. And then went on to say that militants were nothing but "tyrants." Two of his aides referred to them as "ideological criminals" and "the new barbarians."

[2] *New York Times*, May 1, 1969.

Let me provide one final quote from a column by Joseph Alsop. "The country's anger with the SDS black militant goings-on is rising so rapidly that it is now reaching a near apoplectic stage. It is not limited to middle class whites either. Many people could join the coalition of the un-young, the un-black, and the un-poor." [3]

There are two things we can do in the critical situation we now confront. We can get so emotionally upset by it that we are completely involved and cease to be able to think. Alternatively, we can analyze what is happening and try to affect the dynamic within which we are living. I decided a long time ago that a single individual does not affect immediate events. It may look as if he does, but immediate events are controlled by long-run social trends. If you want to bring about significant change, you must work with others to create new trends.

In this discussion I am not going to talk very much about technology, human values, or leisure. Quite honestly, I don't know what any of those words mean. Indeed I don't think any of us can define them effectively because those words are part of an industrial-era language which is obsolete. Until we can create a new language, we cannot discuss our real problem-possibilities. In one sense, therefore, I want to talk about creating a new language.

I think that in historical perspective the great success of the sixties was that we were able to define the overall nature of the crisis in which we are presently living. If America comes to understand it clearly

[3] In November, 1969, a Gallup poll showed that Americans believed that young people were almost as dangerous as Communists to the United States.

enough, I believe that the crisis can be dealt with. I would argue that our central problem is that we have no method of dealing with what I call "system breaks." In other words, when major changes in the technology and culture occur which make obsolete our present ways of thinking, we have no pattern that enables us to control them. (These system breaks are, of course, occurring more and more rapidly because of the growing speed of change.)

In the past the effects of system breaks worked through the system as a result of economic crises, changing leadership, group takeovers through violence, and limited wars. We have now decided that disruptions of these types should be avoided. The elimination of these disruptions, however, ensures the preservation of deadwood throughout the culture: this deadwood is strangling much potential new growth. Joseph Schumpeter showed that economic crises are ways of shaking out the deadwood and allowing the new growth to come through. Shifts in leadership, particularly violent ones, and wars have performed the same function.

Our growing inability and unwillingness to permit localized system breaks to work themselves through is causing a fundamental imbalance throughout the global social economy and culture. The various unresolved system breaks are now compounding each other and threaten to create the last stage of an overall system break. I call this a Toynbean crisis, and mean by this shorthand the type of crises which caused the end of a particular culture in the past. Toynbean crises occur if enough system breaks happen at once and move out of control.

A Toynbean crisis occurs when the socioeconomic system and the culture is so out of phase with reality that it ceases to be effective. I believe the United States is reaching this point. I fear that if it does, it will bring down the rest of the world.

Such apocalyptic visions are no longer fashionable. It is urged that we will join together with the Russians and tensions will therefore decline. I find this point of view ludicrous; if the rich countries join together we move toward a split between the rich and the poor countries, which is inherently more serious for it involves a split between the white and the nonwhite nations of the world.

What must be done? It is clear that two myths must be destroyed before we can even begin to move. The first is the myth of progress, the great myth by which everything that happens in the future is seen as better than what happened in the past, regardless of its nature. In the West, we have believed that we were doing good, whatever we were doing, and indeed this is still the dominant industrial-era myth.

There's a second myth that must be destroyed: the circularity myth. In other words, all we do is go around in a circle: we were tribal, we became detribalized, and now we are retribalizing. But technological tribalism is not the same as nontechnological tribalism. Indeed, technological tribalism will be profoundly dangerous unless we come to understand it better and to prevent the separation inherent in previous tribalisms.

At this point, I must make a choice regarding the nature of what I should say. I have stated that two myths should be abolished: I could now tell you with what they will be replaced. If I do so, you, however, will feel comfortable, for you will recognize the basic university pattern. The "expert" tells you what he believes. You listen and ingurgitate, holding your critical faculties in abeyance.

However, I want you to think, so I'm not going to give you any predigested conclusions; rather I'm going to try to force you to think things through for yourselves. In other words, I am not particularly interested in my *own* conclusions, I know they are inadequate. I know that unless we all participate in creating our future, it is black indeed.[4]

Where do we start? I think we must first recognize that we have to move away from a discipline-centered information structure to a problem-centered information process. This is not something I'm really willing to spend much time on here. If we haven't understood this point yet we really are a long way back. Disciplines are fragmentations of gestalts and totalities: one can only meaningfully analyze totalities. If one breaks down reality into its economic, sociological, psychological aspects, one can discover nothing significant. What we need are new forms of institutes to study problem areas. The job of those institutes should be to discover everything we can learn about those problem-possibility areas.

I could go on for some considerable time about how problem-possibility institutes could be structured, but I'll only name a couple of points.

1. It is essential, at least in the immediate future, that members of the institute do not live in the place where the institute is located. Such a pattern acts to isolate the most vital information on the problem-possibility area at a single point and limits the possibility of diffusion. Instead, we need an institute where the

---

[4] For a view into the future see my book *An Alternative Future for America* (Chicago: Swallow Press, 1968).

staff would be composed of apprentices. I am going back to a very old word here. The experts—or more properly "facilitators"—would come in once every three months or so. They would work together for between three days and a week, set up a working program to operate between their meetings, and leave; the apprentices would then carry out the program.

2. The second requirement is that there must not be a sharp break between the status of the experts and that of the apprentices. In the institutes, as indeed in all education, interchange must be based on sapiential authority instead of structural authority. We must move away from the concept that people have the right to give orders because of their position and toward the belief that people only have a right to give orders when they have the appropriate knowledge.[5]

This should not be a startling idea. I find that young people understand it. They are not asking for an end to authority but are searching for an alternate system of authority. They deny that it is sufficient for a person to answer a challenge to a profound policy with the statement that he wishes it. The only sufficient answer to a challenge to a policy is to demonstrate the reason for the policy: the only occasion when it is justified to avoid an explanation is when there simply isn't time for one.

Our second need is to discover better ways to move information. It is often argued that we live in an information environment—we do not! The sooner we get this point clear, the better off we are going to be. If I may paraphrase: Never has so much information been moved to so many for so little purpose.

[5] I owe this formulation to Tom Patterson.

I developed a definition of an optimist and a pessimist some time ago which I think is relevant. It relates to what people do with the paper that arrives on their desk. We all get more paper than we can read and we therefore create piles of paper. When the piles get so high that they are psychologically threatening, we have two choices: we can simply junk the whole lot—and that's what we do if we are pessimists; alternatively, we can go through the piles putting aside what we really think is important and that can form the basis for the next pile.

I think the answer to this problem of excessive information is new forms of information structuring. One of these on which I have been working is the dialogue-focuser. The first question raised is, What do we already know about this topic? There is usually a good deal more known about the topic than is assumed. Let us take the field I've been most interested in recently—poverty. There is today an agreement that there are poor people in this country. This is an achievement of the 1960's. There is even an essential agreement that there are starving people in America. This is an achievement of the late 1960's. The second point the dialogue-focuser raises is, What issues do we disagree on? What are the things we do not know about? The third question the dialogue-focuser raises is, What do we believe might be the way in which we could solve current disagreement. After the essential information has been correlated to answer these three questions, it can be published at various levels of difficulty.[6]

---

[6] A series of books has been published incorporating the idea of the dialogue-focuser. Called the Dialogue Series and published by Bobbs-Merrill, it includes the following titles: *Dialogue on Poverty; Dialogue on Technology; Dialogue on*

Obviously the statements in a dialogue-focuser cannot be final: rather they must be in a process of continuous revision. However, if they are well produced we can ensure that we don't spend our lives talking about issues that have already been decided, and can thus avoid the situation that exists in almost all our classes at the present time. (In addition, of course, these statements can be produced not only in printed but also in video and audio form.) I think we could create helpful dialogue-focusers for very many subjects in the next two to three years if we would try. However, I see no evidence that we shall do so.

The next step is to stop ruining people's lives by the education we are giving them. Buckminster Fuller was once asked if he was a genius, and he said, "No, of course not. There are no geniuses, some children are less damaged than others." This is, of course, the tragedy in the education issue as we face it today: we are in a double-bind situation. Whatever we do now is wrong. If we fail to suppress violence in schools and universities, we will create enormous tensions among those who don't understand the reasons for the violence. If we do suppress the violence, we take away the symptoms that should be alerting the country to the critical nature of the present crisis.

The time to find out that you don't have enough gas to get across the Atlantic is before the plane is halfway across. Today we are very close to being more than halfway across the Atlantic on many issues; particularly the ecological. Harrison Brown, however, will cover this topic and I don't intend to anticipate his argument.

Education; Dialogue on Women; Dialogue on Science; Dialogue on Youth; and *Dialogue on Violence.*

I shall therefore return to the subject of education. What do we know about education? We have recently developed a new theory which contradicts essentially all our educational practice. First, people will only learn effectively about things which interest them. People learn only when they want to learn. Incidentally, the fact that you get people to put data into their minds temporarily so they can regurgitate it doesn't show anything about learning. After the regurgitation process has been completed, nothing is left. If you retest somebody on the same material four weeks after they were meant to have finished with it, they will remember very little. The second educational reality is that you cannot change anybody's mind directly. This is profoundly disturbing. In other words, if you want to change somebody's mind, it is not effective to challenge something he already thinks he knows. If you are going to help people to learn how to live in their own environment, you have to take a topic on which people do not presently have strong convictions; although they are interested. They must then rethink for themselves the views which they presently hold.

We need to achieve this process of rethinking not only for isolated individuals, but for the whole society. I believe it would be easiest to do this around the issue of ecology. We then need to carry through three steps. First, we need to bring accurate new information into the homes of as many individuals as possible. This would require, of course, the use of network television. Second, we must provide people with the opportunity to discuss this new information so as to discover its relevance to them. Finally, we must show people ways to be involved in action so that they can test out their new understandings through examining the feedback to steps which they believed would be valuable.[7]

Unfortunately, there is no sign as yet of crash courses in reality on the required scale. Let us therefore look at what can be done immediately and on a small scale. First, it is clear that the way to start a young child's education is to answer his "whys" as they develop. Effective answering of whys cannot, however, be achieved in large classes. If you want to find that out, go and discover what happens to five-year-olds as they grow into seven-year-olds.

Second, we must abolish the distinction between teacher and learner, particularly on campus. One way to do this would be to introduce, as fast as possible, middle-aged people into existing learning situations. The major factor which aggravates tensions in the teacher-student relationship is that the teachers are older than the students. If you break this age relationship you begin to change the patterns in class.

Let me sum up what I have said so far. We are moving from the industrial era to the cybernetic era. This leads to many types of system breaks: the systems we have relied upon in the past cease to be appropriate. So far we have dealt poorly with these system breaks. As a result there is a danger that a massive final Toynbean system break will emerge.

We will only learn to handle system breaks by the creation of new institutions in particular areas, the creation of new forms of information structuring and new methods of involving people in their own educa-

[7] For further discussion of these issues see "Communication to Build the Future Environment," developed in connection with a conference of the same name, in *An Alternative Future for America* by Theobald.

tion. Today our overall information-transmission system makes it impossible for people to find out what sort of world they are living in. As an example, let us consider the reporting of educational events. The mass media report the disasters. They ignore the creative patterns, although I am convinced that a small amount of research would come up with many exciting developments on campuses across this country. If the media reported these also, we would then have a balanced picture of what is going on on campuses instead of coming to believe that all students are rebelling.

The media make the news by determining what is newsworthy. It is the decision of the media that violence is newsworthy while creative actions are not, which leads people who wish to communicate to do so through violence. Violence seen on national television has different consequences from those which occur if awareness is confined to the local level. Much of the blame for the escalation of the violence lies with the media.

Today you attract attention by violence. If you are not willing to be violent, it is very difficult to gain attention: the temptation therefore is to gain attention by violence. It is surely inadequate, if not dangerously wrong, to blame those who use violence without admitting that those who control the mass media have so structured the payoffs in the society that only violence can be heard.[8]

[8] I believe, even so, that while violence does get the attention necessary for communication it also prevents any meaningful transfer of information. However, acceptance of this reality requires that one recognize that there are no truly effective methods of mass communication today. While I am ready to accept this as a temporary state, it is hardly surprising that the young are not.

Where do we go from here? The first necessity is to perceive clearly the nature of the crisis. We have a double-bind situation. If we permit violence to continue, we shall anger those in the silent majority. But if we suppress the symptoms of change we shall turn off the coalition of the young, the poor, the minority groups, and the women that alone can hope to convince the country that fundamental change is required.

There would be an enormous cost, both personal and societal, if this latter development should occur. The personal cost would be a complete breakdown of the life-style of the individual. The rising crime rates and growing anomie of the ghetto can best be explained by the absence of meaning and movement in the Negro ghetto. If there isn't anything better to do with your time, why not rob a store, why not take dope, why not permit a catastrophic breakdown in your life? Similarly, if young people can find no causes for which they can struggle meaningfully, they will drop out into a culture of introspection, drugs, and meaningless sex.[9]

It is at this point that each of us faces a dilemma. We must decide whether the industrial era in which we are living can be expected to continue or whether we are all indeed participating in the transition from the industrial to the cybernetic era. If the industrial era is to continue, we should fight against its injustices, as has been the pattern throughout human history, If, however, we are participating in a transition, we are able to leave the past injustices behind and to spend

[9] Between the date this speech was given and its revision, the change of culture suggested as possible has become very visible. People on campuses have become far more tribal, congregating in smaller groups with little concern about overall issues. We are now re-creating a form of tribalism that is totally inappropriate for a technological era.

our time creating a new socioeconomic system based on our growing knowledge and abundance.

It must be clear by now what I believe. We are all living through a transition. If this is the case, we need to develop new techniques to achieve change.

We have to understand the past to dream a future and create the present. We must understand the past because it determines what society is able to do in the present. We must dream about the future, for only if we have new dreams can we analyze in new terms. And we must create the present, moving it in small steps out of a past which is no longer relevant into a future which we desire.

Now let me conclude. What will our future hold if we survive? The first thing we are going to have to understand in our new world view is that each individual is unique. This implies that equality is impossible. We can only aim to achieve the self-development of each individual to the maximum of his potential: this will not create equality.

Second, unique individuals can only develop and enjoy their uniqueness within diverse communities in which these uniquenesses can be honored.

Third, we have to recognize that mankind and, in fact, all organic systems are self-actualizing; in other words, there is a pull into the future. It is impossible to explain actions except on the basis of a pull into the future. It follows that one must be involved if one is to be relevant; it is impossible to be objective and effective.

Fourth, real change occurs as a result of process. Things don't happen all at once. Unfortunately, Americans like nice, neat, tidy, immediate solutions. As

Margaret Mead once said, "In America the long run is six months."

If, as I believe, we are moving into a new world, we must accept the fact that we know very little about it. We are all joint learners. Let me therefore conclude by inviting all of you to become involved—to share in the creation of a desirable future. This is a new profession; it is hardly overcrowded! All you need to belong are a passionate concern for the future of man, a willingness to admit your ignorance, and the courage to act when action is required.[10]

---

[10] A lengthy discussion of the shape of a possible new world is contained in *Teg's 1994*, created by my wife and myself. This is a participation book designed so that the reader can discover for himself what he believes about the next twenty-five years. Copies can be obtained from Personalized Secretarial Service, 5045 N. 12th Street, Phoenix, Arizona 85012.

EMMANUEL G. MESTHENE

TECHNOLOGY AND HUMANISTIC VALUES [1]

        What I should like to
talk about is the relationship between technology and
the humanities. In doing so, I shall be construing both
terms very broadly. I will mean by "technology" the
totality of the tools that men make and use to make
and do things with. Our technology, then, is our society's tool box, which includes not only hand tools and
machines, but also the spectrum of intellectual tools,
from language to ideas to science, and to such latter-
day techniques as computer programs, systems analysis,
and program planning and budgeting systems. It is
with this tool box that our society and the people in it

---

[1] This chapter first appeared in the October 1969 issue of
Science Journal, published in London. Reprinted by permission.

do their work; the nature of work cannot be understood apart from the concept of tools.

When construed in this broad way, of course, the concept of technology begins to shade into the wider concept of knowledge, which is why we often hear our time referred to as a "knowledge" society. There are a number of important issues, in fact, that have very little connection with technology as such, but are nevertheless relevant to technology by virtue of their illuminating of the social role of knowledge in general. Knowledge in this general sense also includes information of all sorts, intellectual methodologies of all sorts —such as the use of computers by the arts and humanistic disciplines, for example—and extends further to a commitment to the value of rationality, and to the multiplication and growth in influence of a host of knowledge institutions, from universities to research and development institutes to analysis and planning staffs in public and private organizations.

It is through enhancing the status, office, and importance of knowledge in one or another of its forms, in other words, that science and technology may be affecting society most significantly. In that sense, understanding technology in the broader sense of knowledge—in the ancient sense of *scientia,* if you will—may be a precondition of understanding the relationship between technology and the humanities, or any other aspect of society that we may be interested in.

I construe the word humanities equally broadly, as meaning the concern with the expressive, moral, and contemplative aspects of living as distinct from the instrumental aspects. The arts are the concern of the humanities in this sense, as are history, and philosophy

—at least in its original sense—and our aesthetic, ethical, and religious values. It is to this side of life that we appeal when we speak of the value of leisure —but leisure in the best sense, not in the potentially self-defeating sense of time off from work. The philosopher Thomas F. Green has recently expressed that distinction well:

> Leisure is not opposed to work in the same way as free-time is opposed to work-time . . . nor does the conception of leisure derive its meaning from the conception of work. We have instead a view according to which the idea of leisure has meaning in its own right and cannot be interpreted in relation to the idea of work. It is an understanding of leisure that is a wholly different plane from that of free time. It is an ideal, a fundamental one, having to do with the totality of a man's life and character. It may be, as the ancients thought, the highest ideal of a civilized society.[2]

Just as the nature of work cannot be understood apart from the concept of tools, in other words, so cannot the nature of leisure be understood apart from the idea of expressiveness and appreciation as human activities that are valuable in their own right.

You will note my emphasis on leisure and contemplation as expressive activities, in contrast to instrumental activities. The distinction can be made in a number of ways. It is the distinction between knowing, on the one hand, and understanding on the other; between analysis and synthesis, facts and meanings, answering and questing, researching and thinking, learning and interpreting, doing and appreciating. The domain of

[2] *Work, Leisure, and the American Schools* (New York: Random House, 1968), p. 72.

the humanities, as I understand the term, is to be concerned with the second terms of all such pairs.

What, now, is the effect of technology on that domain? How are the humanistic disciplines dealing with the spectacular elaboration and enrichment of our society's tool box? Put another way, with what challenges does modern technology confront our humanists, our artists, and our theologians? I want to suggest to you that the challenges are significant, that they are poorly understood and poorly met, when they are met at all.

One of the problems is rooted in the seductiveness of fine tools. As everyone knows who has worked with tools—whether mechanical or intellectual—it is a pleasure to work with tools when they are suited to the job and a drudgery to attempt a job for which they are inadequate. There is therefore a built-in bias in favor of undertaking jobs for which appropriate tools exist, to the neglect of jobs—however important otherwise—for which the right tools are not available. That bias then takes the form of a tendency to define the job in terms of the tools; that is, to identify what needs to be done with what can be done. This tendency is discernible far more widely than just in the humanities, of course.

It is discernible in our national objectives, which are disproportionately weighted in favor of defense and economic growth, because our aerospace technologies are so effective and because a constantly rising GNP is easier to measure than accomplishment in more intractable areas. As the noted economist Wassily Leontief noted recently:

In economic growthmanship the method tends to become the goal. And if some of the human consequences of this condition are not pleasant to contemplate, if the "external costs" of growth clearly seem to pose dangers to the quality of life, there is as yet no discernible tendency among economists or economic managers to divert their attention from the single-minded pursuit of economic growth.[3]

The tendency to define the task in terms of the tool is discernible in many other areas of public policy also. The techniques that we group under the general rubric of systems analysis have had a modicum of success—though much less than their practitioners would have you believe—in the military and aerospace fields, where the goals are clear and political constraints are minimal. We have been too ready to assume—wrongly as we are beginning to discover—that they could be equally helpful in education, whose goals we are neither clear about nor in agreement upon among ourselves, or in resolving the problems of our cities, whose essence lies in political and economic conflicts.

The social and life sciences are subject to the same reductionist tendency. Economic theory postulates a system of small producers of consumer goods and services in competition with each other and catering to the sovereign tastes of an independent public. The reality, as we know, is industrial concentration, manipulation of demand, and a growing need for social goods and services, but the otherwise sharp tools of economic analysis are inadequate to the reality so that economic methodology increasingly constrains the subject matter of that discipline. The tendency is to be found also in a sociology become so empiricistic that it has virtually abandoned all concern with social values; in a biology that defines the organism by the concepts of

[3] *New York Review of Books*, October 10, 1968, p. 35.

physics and chemistry rather than by the integrity—Aristotle called it the entelechy—of its object; and in a psychology that still concentrates much of its attention on the relatively easy task of describing behavior rather than the much more difficult one of discovering its causes.

The same reductionist error is found in the arts and the humanities. There is surely nothing wrong with the exploration of new expressive possibilities inherent in new technology. Architects have always done this, as construction technology has moved from the fitting together of marble slabs, to the Gothic arch, to steel and reinforced concrete. The invention of the photographic and motion-picture cameras extended the range of the visual arts, and the modern piano and large sound box were necessary conditions to the arts of Beethoven and of Heifetz.

None of this is in dispute. My concern—to which I will be devoting the last part of my discussion—is with the artist who sees in technology only a new instrumentality; that is, who sees its artistic possibilities but not its aesthetic potential. My worry about the new breed of joint societies of painters and engineers or composers and electronics experts is not with the fact that experimentation along such lines is undesirable, but with the too-ready conclusion that such experiments exhaust the artist's responsibility to deal with technology. There is much more to the matter than that. There is the task of appreciation in which the artist should lead.

I have had in mind also, in the preparation of these remarks, the current uses of computers in humanities research. Again, there is nothing wrong with doing content analyses or concordances in one year instead of in fifty. The danger is that the dazzling efficiency of it all will make content analyses and concordances appear to be the most important work that humanists can do. Similarly by the way, with historians of technology who write volumes of descriptions about how machinery works and forget to trace its social impact, and with philosophers who are so seduced by their sharpened instruments of analysis that they forget the responsibility imposed on them by their calling to engage also in social criticism and disciplined speculation.

This seductiveness of fine tools that I have been describing—it becomes a seductiveness of rigor in mathematics and in the intellectual disciplines that can find uses for mathematics or for computers—must be accounted one of the costs of the scientific revolution that began a little over three hundred years ago. Understandably, modern science took off and thrived on exploration of the easiest possible subject matters—the inanimate ones. A condition of success for physical science, moreover, is identification and investigation of the easiest possible problem. No physicist or chemist will tackle a difficult problem—however exciting in prospect—if he can handle an easier one first. And the subject matter—physical nature—allows this procedure because its secrets are in the form of cause-effect connections, i.e., instrumental relationships. The order of discovery is therefore irrelevant to its objective, and can be left to the determination of available tools without fear of distortion.

The social scientists, by contrast, must be mixed in their objectives. It is certainly possible to concentrate only on the cause-effect relationships in the situation, as so many of the social sciences do at the present time. But in thus seeking to capitalize on the methodological strengths of the physical sciences, they must necessarily impoverish a subject matter that can never be fully understood unless its cause-effect connections are seen in the context of the purposes and organic relationships that are an inescapable part of it. Just how such seeing in context is done and what it purports for the formulation of knowledge about social situations are the difficult and essential objectives of the social disciplines, which are short-changed so long as the methodological presuppositions of the physical sciences are taken to be universal for all inquiry. Measurement is an adequate tool only for the measurable aspects of a subject matter. To define it, then, in terms of what can be measured is to distort it.

When the principal concern of a discipline is the expressive, moral, and contemplative—as it is with the humanities—the danger of abdication is greatest. There is a scientific aspect to these disciplines too, of course; writing has its grammar, music its acoustics, architecture its engineering, and philosophy its logic. But this is clearly not the essence of these disciplines. Their essence is to explore possibility, to discern meaning, to elicit value, and to elaborate ideal vision in the world in which they function. The world in which they function in our society is a highly technological world. If their only response to technology is to explore it for finer tools, they fail in that which we most demand of them; i.e., that they lead us to see—in the Platonic sense of seeing—what it is that technology imports for man: for his aspirations, his values, and his gods.

This is the real challenge of technology to the humanities, and it is a challenge that is being only indifferently met, as I have said. That is my principal theme, which I should now like to develop somewhat.

The glory of literature and art, I take it, is that they help us along on an imaginative exploration of possibility. This point has been made well, I think, by the philosopher John Herman Randall, Jr., who says that the work of the artist

teaches us how to become more aware both of what is and of what might be. . . . It shows us how to discern unsuspected qualities in the world encountered, latent powers and possibilities there resident. . . . Art is an enterprise in which the world and man are most genuinely cooperative, and in which the working together of natural materials and powers and of human techniques and vision is most clearly creative of new qualities and powers.[4]

Then Randall goes on with this interesting comment:

As a discoverer of new powers and possibilities [the painter or the poet] has much in common with the scientist. What he does with and makes out of what forces itself upon his attention and what he sees, by selecting from it, manipulating it, reorganizing and restructuring it by means of his distinctive art, is very much like what the experimentation of the scientist effects.[5]

The two quotations taken together suggest the natural intersection of technology and art. Technology is in the end an enterprise of making possible what was not possible before, as, for example, it was not possible before rocket technology to go to the moon or to communicate by satellite. Science, which is comprehended by technology in the broad sense in which I am construing it, also creates new possibilities: new attitudes and perceptions, new belief systems, new ways of approaching and dealing with the world.

[4] *The Role of Knowledge in Western Religion* (Boston: Beacon Press, 1958), p. 128.
[5] *Ibid.*, p. 131.

The artist—best equipped by temperament and training to discern and explore possibility as such—would, one would think, revel in this richness of new possibility that technology produces. He would find new worlds for his vision to penetrate, new challenges in teaching the rest of us to see as well. Yet too many contemporary artists stand angry and uncomprehending before the new possibilities that technology creates, and they produce poetic and dramatic complaints about a world in which they see only threats, or literary paeans to the wonderful time before the Bomb. Elizabeth Sewell, in a typical example, speaks of "the enormities with which we have been confronted by science and technology within our lifetime. . . . The first is Auschwitz, . . . this first terminal point of our technological age. The second terminal point is Hiroshima." [6]

But technology is not a modern invention. The trireme and the guillotine were technological devices too, but Homer and Victor Hugo did more than lament about them; they could condemn terror with the best of them, but they also could evoke the wonder and awesome possibilities of revolutionary times. What distinguished the music of Beethoven from that of his predecessors, to use another example, was its infusion with and evocation of that same revolutionary change, not the development of new musical technologies—unless you would equate the cannon in the Battle Symphony with the electronic devices with which some modern composers are experimenting.

I repeat that I find nothing wrong with artistic experimentation with new devices, or with the use of computers for establishing concordances. What I have

[6] "Science and Literature," *Commonweal*, May 13, 1966, p. 219.

yet to see is the artist able to take the full measure of our technological times. And that is a failure of our artists, not of our technology.

It is incumbent upon me, I suppose, to go beyond exhortation and suggest somewhat more concretely what I have in mind. I shall try to do so in the context of some concluding remarks about religion. I choose religion for three reasons. First, I find a failure of theology and of our religious institutions analogous to the failure of our artists, and for many of the same reasons. Second, I have myself thought more about the problem in relation to religion than to art, although I find a very close kinship between religious experience and aesthetic experience.

Finally, I choose religion on a hunch that it may contain a key to the problem of leisure—in the classic sense—and to the humanistic need for contemplation that I alluded to earlier. Those who worry about the need for and the role of leisure in contemporary society stumble at every turn over the problem of passivity. Succeed as we might in improving the quality of television, or proliferating community art centers, or establishing a chamber music group in every town, we are left still with the nagging feeling that all this is palliative, that we miss the involvement we seek because it all lacks reality somehow, because it is not really leisure we are dealing with, but the filling up of free time, to allude once more to Thomas Green's distinction.

Rather must leisure be *sui generis;* not only an activity engaged in, in its own right, but the greatest, the most essentially human activity. The essence of that activity, as I have tried to suggest, is expression, contemplation, imaginative vision. That is why it is so necessary to get the arts and the humanities and religion on the right track again—and ultimately religion. I suspect, for it is in identification with transcendent meaning that we may find true leisure, and with it true humanity.

Religion is not immune to the impact of technology. The function of religion is to direct man's eyes to a vision of the eternal. Once again—to put it in the terms of my present discourse—its function is to contemplate the realm of possibility that transcends the actual and helps to give it meaning. Technology, as I have indicated, transforms possibility by adding to it. The major effect of technology on religion, therefore, is to render inadequate those formulations of the eternal that are based on an earlier experience; it undermines historical conceptions of God. One would think this might challenge the modern divine to explicate the nature of God in the context of contemporary experience and thereby enrich the lives of men. Some are beginning to do that, but most men of the cloth today seem content to function as social workers, an occupation which is worthy enough in its own right, no doubt, but is hardly the heart of the calling.

What, then, does modern technology require by way of religious commitment?

I begin with the assumption that there is a distinctively religious dimension to human experience that is related to but distinct from physical, social, psychological, intellectual, or strictly aesthetic experience. I infer from this that the object of religious experience is distinct from the objects that are involved in the other kinds of experience.

An object adequate to religious experience must, I think, have intellectual, moral, aesthetic, and emotional dimensions: that is, it must be somehow consistent with what we know; it must coincide with what we value; it must accord with how we see the world; and it must embody our sense of the unity and therefore the meaningfulness of human experience. But the functioning of the religious object as such is not constrained by what it is specifically that we know, or value, or see, or feel at any given time. It is independent of particular contents—and is therefore eternal—because what it does, rather, is help us celebrate and glory in the fact that man can know and value and see and find meaning in his experience. Only man, so far as we know, has this need and capacity to revel in the ideal possibilities that are implicit in his own nature; it has been called the divine spark in man. And since celebration, glorification, and ritual revelry are basically social activities, men act religiously—that is, they share a religious experience—when they join together to worship a common vision of what man might be ideally if he could weaken or sever his earthly bond and transcend himself. That vision of ideal possibility, when clothed in mythological language appropriate to their particular experiences, is what men call God.

If we now attend more closely to moral—as distinct from social and religious—values, we find a similar process of idealization taking place. The sociologist sees values as the generalized set of categories according to which the behavior of groups and individuals in a society can be understood. He investigates the "ought" structure of society, but only as an aid to understanding the moral and legal norms that guide action in concrete situations. The sociologist's concern with values is thus principally descriptive.

The moral philosopher's concern, by contrast, is prescriptive. He is interested in discovering what a society's or individual's "ought" structure *should* be, not only what it is. He seeks to formulate generalized ideas of good and evil, to use them as criteria for normative judgments of behavior, and to offer them as guides to moral action. Just as social values are abstracted from concrete choice behaviors, in other words, so are moral or ethical values abstracted from the imperatives of specifically moral choices. If we now engage in a further abstraction from ethical values at the moral philosopher's level, we arrive at a higher-level generalization of good and evil in some ultimate sense of those terms.

It is at this level and in this sense that values are of interest, in the first instance, to the metaphysician, that is, to the technical philosopher who investigates the nature of existence as such. But it is also at this level and in this sense that values begin to concern the theologian. The metaphysician deals with ultimate good and evil as abstractions. The theologian is interested also in having the sense of ultimate good and evil infuse actual religious experience. It cannot do that, however, so long as these remain pure abstractions, any more than the idealized vision of human possibility that I mentioned a moment ago can function as a proper object of religious experience so long as it remains only a vision. Both must be clothed in the language of myth.

It is in religious myths, then, that both the vision of the divine and the sense of ultimate value take on a form which enables them to function in religious experience. The form, as in all myths, is that of a story, an idealized or imaginative narrative. In its form, therefore, a religious myth is primarily a work of artistic creation—an aesthetic object—which is perfectly appropriate when you recall that religious experience is first of all an experience of celebration and glorification—of enjoyment, if you will, of perfectability and value as ideal possibilities.

Particular religious myths and religious values are, of course, dependent on the social and moral values of any given time, but the value of religious myth and the function of religious values are not so dependent. Since social values change under the impact of technological change, therefore, the content of the myths in which they find their religious form must also change. Religious mythology and codified morality need to evolve as the values of a changing society evolve. If they do not succeed in doing so, they will leave a vacuum that some secular myth or some nihilistic antimorality will seek to fill—either some super-rationalistic scientism of the sort that writers like Jacques Ellul keep warning us about or a what's-the-value-of-values philosophy such as is characteristic of a segment of modern youth.

To return briefly to art, and to a point I suggested earlier, I am sure many of you have realized now that my treatment of religious values sees them as closely akin to aesthetic values, and it is in exploring the influence of modern technology on aesthetic values, I suspect, that we may gain our most important insight into how tools—technologies—influence humanistic values. It is important to recall that every artistic process is a process of "making," and that an instrument or tool—or technology—is essential to any process of making.

What is less easily seen is that aesthetic appreciation, that is, the value, of the completed art object cannot be independent of the process, and therefore of the means or tools that were used to make it. Analogously, when we have looked at technology and the science that underlies it directly, we have tended to concentrate so exclusively on the process and the tool that we have neglected consideration of the ends to which they have served as means. This partial view of both art and science accounts for our traditional view of them as separate activities.

A major consequence of the prevalence and power of our modern technologies, however, is that we begin to realize that the ends—the probable social consequences —must henceforth inform the process of science and technology, since the cost of treating science and technology as ends in themselves is becoming too high. In different words, as our knowledge and technical capability grow, their inherent value takes a lower place in our system of values relative to other values that they serve. The value of the means lessens in the perspective of the value of the end. That suggests that the process of knowing—the technological process, in the broadest sense—is not so different from the typical artistic process as we may have been brought up to think.

The hypothesis to which I am led by these considerations is that there is a sense in which all values are ultimately aesthetic—having to do with relationships of means and ends and deriving from man as creator; that is, from man as maker, whether he is making a painting, a rocket, a book, a polity, a good life, or his god. If there is substance to this hypothesis, and if this hypothesis can lead to a conception of a truly human view of the nature of leisure as contemplation, it will be a careful pondering by humanists on the nature and operation of technology that will lead to that conception.

HARRISON BROWN

TECHNOLOGY AND WHERE WE ARE

I believe that the most
overwhelming characteristic of the world today is the
rate of cultural, social, and overall environmental
change which is being brought about by an accelerating
rate of scientific and technological development. The
rate of change today is far greater than it has been in
the entire two-million-or-so-odd years of human existence, and we have every reason to believe that this
rate of change itself is accelerating. But before we look
into the future, I believe it is important that we attempt to understand what the main technological developments were which brought us to our current condition.

Concerning change, I would like to tell a story about
a New York businessman who went to Switzerland on

a holiday to do some mountain climbing. As he was walking he suddenly felt rather tired, and he crawled into a cave and fell asleep. He woke up and was dismayed to notice that he had a long beard. He immediately went to the village where he'd been staying, for, if what he suspected was true, he had slept for twenty years. He said to himself, "Good Lord, what about my stocks!" He immediately placed a telephone call to his broker in New York. The stockbroker's voice was rather feeble by then: yes, he did remember him; he would look for the portfolio; he found it; it was dusty; and he said, "Now let's see, you owned a thousand shares of General Motors. That's been split five times, so that's now worth 286 million dollars. You owned fifteen hundred shares of Chrysler; that's been split eight times, and it's worth 400 million dollars now." He went on and on, and suddenly the man who had been asleep realized he was a billionaire. He was ecstatic. Just at that moment the operator came on and said, "Your three minutes are up, would you please put in another one and one half million dollars?"

Manlike creatures have existed on this earth for something like two million years. During the greater part of that time they have lived pretty much like the animals about them. Occasionally a technological innovation came to the forefront making life somewhat easier, and also making it possible for human beings to spread. The controlled use of fire made it possible for mankind to live in climates that would otherwise have been uninhabitable and extended the range of food which could be eaten. Clothing made it possible for mankind to spread over the surface of the earth. He got his food by hunting it and by gathering it, and so, even when man had spread over the entire surface

of the earth, the population of human beings at satura-
tion was never more than about ten million persons,
about the population of the greater New York area
today. And that is the way people lived for virtually
two million years.

About ten thousand years ago, there was a tremendous technological breakthrough. Man learned that
he could nurture those plants which were particularly
useful to him; he learned further that he could nurture
those animals that were particularly useful to him and
that he could destroy those plants and animals that
got in his way. This new technology had a profound
effect, for it made it possible for literally thousands of
human beings to be supported on land which had previously supported only one person, giving a potential of a
thousandfold increase in human population. The invention of agriculture had another important effect:
it turned out it was possible for a man and his family
to grow somewhat more food than they needed just to
support themselves. There was a slight surplus. This
surplus was never large, it never, until very recently,
amounted to more than about 10 percent of the total.
But it meant that about 10 percent of the human population did not have to scrounge for food. Somebody
would provide the food for them, making it possible
for them to do other things.

And so, directly associated with the invention and
the spread of agriculture, we find the first emergence
of professions; some persons became soldiers, some became architects, some became scientists, and so forth.
Ten percent of the human population could live in this
new way. Directly associated with these changes was
the emergence of cities and of the great ancient civilizations in the valleys of the Tigris and Euphrates, the

Nile, and the Yellow River. This kind of living was amplified and permitted to spread more rapidly as a result of another technological innovation, namely, the use of metals. The use of copper became widespread, making it possible for human beings to do things that they could not do otherwise. However, the use of metals was confined to the cities and to the wealthy class. The abundance of copper in nature is rather low and, as a result, it was very expensive. Elaborate trade routes were established in order to get high-grade copper ore. The Egyptians established a vast combination of land and sea trade routes for copper, taking it all the way to Northern Europe. Iron is much more abundant than copper, but unfortunately the technology for extracting metallic iron from the ores is rather complicated, requiring furnaces that permit a much higher temperature than is needed in the reduction of copper. As a result, iron technology remained unknown for many, many centuries.

Finally iron technology did emerge. Iron is hundreds of times more easily available than is copper, and, directly associated with the development of iron technology, we find the spread of agriculture to Europe, to heavily forested regions which prior to that time were simply not available to the farmer. In this way, iron gave a boost to the capacity of the world to support human beings, and a new population explosion was underway. The United Kingdom was in a particularly favorable position at that time because they had a lot of iron ore and a lot of trees. The way one makes metallic iron is to cut down trees, make charcoal out of the trees, then use it to reduce iron ore to metal.

So the British became a major exporter of metallic iron. This was all fine until they began to run out of trees. Indeed, this was a very serious crisis, for England came just to the edge of having to close down completely its iron-production facilities. One family, the Darby family, supported on its own a research and development program for two generations—and just in time. There was considerable coal in England, but the coal had impurities in it, and, as a result, the charcoal was not usable for reducing iron ore. The Darby family learned how to treat coal in such a way that it could be used in iron manufacture. This permitted England again to go off to the races, because England had a great deal of coal. And so this one technological breakthrough made it possible for England to become a major manufacturer of iron ore again.

One of the new difficulties in England soon appeared in the form of coal-mining technology. The British very quickly used the easily accessible coal, even though they had to dig very deeply for it. In the course of digging they encountered groundwater, which meant they had to pump the water out. This required manpower, horsepower, or something to run the pumps. One day, a genius thought of hooking up a newly invented steam engine for the purpose of pumping out the water. The primitive steam engine soon was developed further, in the form of Watts's engine. Then someone thought of attaching the engine to a spinning wheel; and then someone thought of attaching it to machinery which could weave textiles. The technological history of that period is a marvelous illustration of what we now call feedback, where one technological discovery will precipitate another, which in turn will precipitate another, and will "feed back" into the whole technological fabric of the society. Indeed, at that time, invention followed invention with breathtaking rapidity, and steam very quickly began to replace animal labor.

There is a marked interrelationship between technological change and human values and attitudes. One of the most outstanding examples of that involves the institution of slavery. Slavery was perpetuated as an institution in Europe far beyond the time it should have vanished, had it not been for a glaring gap in our technical knowledge of horse and cow anatomy. Our knowledge of the anatomy of the horse was so poor that we didn't design harnesses properly, and the horse could pull only four times as much as a man. However, a horse ate four times as much as a man, so the hard-nosed businessman of the time would rather use a man any day; a man lives longer, and, though the initial capital investment might be a little higher, he's more amenable and can be trained to do more things. Indeed, this situation resulted in a process of rationalization which perpetuated slavery. Most work that required the concentration of huge quantities of energy was done by mobilizing gangs of slaves.

With the emergence of knowledge of animal anatomy and the development of adequately harnessed horses we see the crumbling of the institution of slavery, except in specialized situations such as those which developed in the new continents of North and South America.

If I had been asked by President Lincoln in the 1860's to make a forecast of what the horse population of the United States would be in 1950 I would have looked at the curves, and noted that there was one horse or mule for every four people. I would have estimated that perhaps the population of the United

States would go up 200 million, so there would be 50 million horses. Actually, today there are only about a million horses on farms. It is interesting how a relatively simple technological development can change a culture.

Our ancestors didn't use horses; they used oxen because they had wooden plows, and in order to get such a plow into the soil they needed a big animal. The ox could do it and the horse couldn't. The invention of the steel- or iron-tip plow, which was self-lubricating, made it possible for a smaller animal to pull the plow. Once the plow was dug in, the horse could move faster and could plow much more land than an ox. One simple invention changed the ox culture to a horse culture, and this had a great effect on the general way in which people lived.

We now find ourselves in a situation where modern civilization can be best characterized as a high-energy civilization. We consume fantastic quantities of energy just to support a single individual. Within the United States today we consume energy equal to the burning of ten tons of coal each year for every person. We also require huge quantities of materials, on the order of 300 pounds of copper, 300 pounds of nickel, and 400 pounds of lead per person. Associated with this upsurge in material and energy requirements, we see a steadily increasing productivity. This has been obvious on the farm, where the production per man per hour has steadily gone up for many decades. We see this happening in industry today, for productivity per hour of labor is going up at a rate which is more rapid than the rate of increase of human demand. We can ask ourselves how far this will carry us. How high can productivity go? Are there any limits? When we look at the situation we must conclude that there are no limits and that, in principle, we will approach this state very shortly—if we don't blow ourselves up in the meantime.

We will soon approach a state where virtually all of the necessities of life are produced by machines with no human intervention. Now, of course, people will be required to design machines and to oversee them in some way, but the actual necessary human intervention will dwindle into insignificance in the future. This is a fact of life we must recognize already in the United States. Some of our economists call us a "service society," now that we have more people engaged in producing services than in producing goods. We can expect this to happen more and more. We will find ourselves in a situation where we will have machines running like mad producing the necessities of life, and, at the other extreme, people will be making nice handmade ashtrays, handmade furniture, and what have you—esoteric luxuries of life.

I don't know how we're going to handle this situation from any purely economic point of view. It's rather interesting to ask our science fiction authors, many of whom are highly imaginative. One of my favorite authors, Frederick Pohl, wrote an interesting story about the ever-expanding economy. We talk about how our economy must always grow, and we ask where that will carry us in the long run. The story starts in the morning; you see a man getting up to go to work. As the alarm goes off the man jumps out of bed, and the first thing you notice is that he is very fat; the second thing you notice is that he lives in a huge house filled with all sorts of things. Then you find that his "work" consists of running out and spending a certain number of ration stamps he is given each month. If he doesn't succeed in spending them, he is penalized by being given extra ration stamps to spend the next month. Then you find that this man's wife complains about having to live in the "slums" and it turns out that the "slums" consist of very large houses where the poor people live. As you go up in the social scale you don't have to spend so many stamps and are permitted to live in smaller houses. Finally, the members of the elite are allowed to live in simple country cottages and are permitted to grow their own vegetables.

This might sound silly, but perhaps it isn't. I believe that we are going to have to give a great deal of thought to the problem of where we are heading in the distribution of the unlimited quantity that our machines can produce. How are we going to eliminate unemployment when no one has to work? Who is going to "own" things? But I cannot end this without putting it within another context, and that is the world context. What is happening in the West today is a phenomenon that results from the harnessing of research and development to industry; we are producing an explosion.

Per capita wealth, although it is very poorly distributed, is increasing very rapidly in the more technologically advanced nations. But that has not happened in the greater part of the world; technology has been introduced in a one-sided way. The economies of the poorer nations are growing at about the same rate as the economies of the richer ones, but, as a result of the one-sided application of our technology, the population growth eats greatly into the economic growth.

Today, the world is dividing into two parts. In one part, per capita wealth and productivity are increasing with enormous rapidity. On the other hand, in the greater part of the world people get hungrier every day and, on a per capita basis, the life of the average person is miserable. This is a situation which is extremely unstable, particularly when we look at it in the context of military power, because our science and technology have placed in the hands of rulers of nations weapons of unprecedented power.

The amount of effort we devote to the production of arms in the world is exceeded by no other area of activity. Because of this I suspect that discussions of what we're going to do with our leisure time in the future are not terribly meaningful. If I were a cosmic gambler I would bet that on the basis of the ways in which people now behave our days are limited. There won't be any leisure, because there won't be any world to have leisure in. It could be, however, that we will come to our senses and do something about it. And here I'm absolutely convinced that our science and our technology have placed in our hands the power to create a world in which everyone has the opportunity of leading a meaningful life, a life divorced from starvation, deprivation, and misery.

Thus far we've not shown any real inclination to create such a situation, to mobilize that power. Just compare seventy billion dollars a year for a defense budget against a dwindling two billion dollars a year for foreign aid. We spend thirty-five times as much money building up armaments as we do helping our fellow human beings overseas. I won't even try to compare our defense budget with the piddling efforts that we're making to eliminate hunger and poverty within our own country. Nevertheless, we have that power, and perhaps we will come to our senses, mobilize it, and create that kind of world.

IMPLICATIONS OF LEISURE

ARTHUR SCHLESINGER, JR.

IMPLICATIONS FOR GOVERNMENT

Technology, human values, and leisure are appropriately connected; for nothing defines our age more than the steady acceleration in the rate of technological change, the consequent production of leisure, and the consequent double strain on traditional habits and values.

In the industrialized world incessant change has become the salient feature of modern life. The world, after all, has altered more in the last century than in the thousand years preceding; and, as scientific and technological changes have acquired cumulative momentum, the pace of innovation increases at an exponential rate. It is chastening to contemplate the statistic often cited by our scientific colleagues, that of all the scientists who have ever lived in the history of the world 90 percent are alive today.

A simple illustration suggests how history has speeded up in our own time. How far can a man travel in three hours? For 98 percent of the long span of human existence, man could travel only as far as he could walk in three hours, say nine or ten miles. With the domestication of the horse, he could go up to perhaps 25 or 30 miles; with the invention of the railroad a little more than a century ago, a moment in the time of human existence, man in three hours could go 150 to 200 miles; with the invention of the airplane, he can go up to 6,000 miles in three hours; and today, with manned space vehicles, man can travel 75,000 miles in the same three hours. Indeed, our life-span has expanded to the point that someone who saw the Wright brothers soar for a few seconds in the air over Kitty Hawk 65 years ago might also have watched Apollo 11 go to the moon.

Nor will this tremendously accelerating pace of change slow up. On the contrary, the driving force of the technological explosion comes from what Alfred North Whitehead called "the greatest invention of the 19th Century, the invention of the method of invention." We are only at the threshold of the transformation. The new age will work in our life, our culture, and our values. The velocity of history has never been greater, and the momentum of technological change is now whirling us past the mechanical age, which began with the industrial revolution of the eighteenth century, and into the electronic age. The world we are about to enter will be dominated by electronic means of storing, arranging, correcting, and using data, and by the subsequent capacity now made available to man to deal instantaneously and precisely with a vast number of previously incalculable factors.

We can only peer dimly into this world that lies ahead. We cannot now visualize in detail the impact of the electronic process on human life and thought; but we can, I think, note tendencies already visible— tendencies that will undoubtedly be accentuated in the years ahead—that will affect the modes of social organization, the character of social disquietude, the quality of human perception, and the very nature of work itself. Let us glance briefly at each of these tendencies because they all affect what will be one of the distinctive problems of the electronic age, the problem of leisure.

The high-technology society which we are entering has already produced striking changes, I think, in the modes of social organization. The high-technology society is, above all, a society of great organization. In advanced industrial states the great organization becomes the unit of social life: great organizations of government, of industry, of labor, of education, of research, of communications. We live in a world dominated by these vast, powering, impersonal, human structures, impervious to individual desire or need; and this is true for all advanced industrial nations, regardless of a system of ideology or the system of ownership. These great organizations order and absorb our existence. And, as they expand, they succumb to what Professor Galbraith has called the "striking tendency of our time, the tendency for organization in an age of organization to develop a life and purpose and truth of its own." This holds for all great bureaucracies, whether public or private. What is done and what is believed are first and naturally what serve the goal of the bureaucracy itself.

The rise of the organization as the dominating unit in advanced industrial society has produced another significant change, and that is the quality of social disquietude or social anxiety. Mankind has never lived in a more unstable environment than this contemporary world, for scientific and technological change incessantly and rapidly renders old ideas and old institutions obsolete. Work itself has played an essential role in the maintenance of personal equilibrium; and the prospective disappearance in the electronic age of traditional forms of work may leave men and women disoriented and adrift in an age where bearings are already hard to fix. Edward Shils and other sociologists are right in stressing that the scope of individual choice and action is greater today than ever before. But the individual, as he develops a keener sense of his own individuality, also becomes more conscious of the threats to this new sense of individuality. For most of human history man acquiesced in powerlessness as the condition of life. He does so no longer, and this makes his resentments all the more acute. As de Tocqueville observed of the conditions before the French Revolution, "A grievance patiently endured, so long as it seemed beyond redress, comes to appear intolerable once a possibility of removing it crosses men's minds. For the mere fact that certain abuses have been remedied draws attention to the others and they now appear more galling. People may suffer less, but their capacity to suffer is heightened." This is why the very strengthening of the sense of individual capability produces a more acute sense of individual powerlessness than prevailed at a time when the individual was genuinely more powerless.

At just this moment, when the individual is beginning for the first time in history to feel a widespread sense of capability, he is confronted, on one hand, by the rise of the great organization and, on the other, by the conquest of life by modern technologies, depersonalizing, mechanizing, and now automating the processes of existence. So the distinctive conflict of our age is not the conflict between nations or the conflict between classes or even the conflict between races or generations, though all these conflicts go on. The basic struggle in modern life is between the individual and the structure. It springs from the desire of the individual to affirm and verify his own sense of identity. Since the young are the most resistant to the discipline of the great organizations, it is not surprising that the form the struggle between the individual and the structure takes in many countries today is the revolt of the student against the institution nearest to him, the university.

Another consequence of the high-technology society must be noted, that is, the impact of the new electronic world on the way we perceive things, on the very structure of perception itself. One does not have to be a devout McLuhanite to heed Marshall McLuhan's emphasis on the fact that this is the first generation to grow up in the electronic epic. The instantaneous world of electric informational media is beginning to alter the way people perceive their experience. Where the preceding culture, the culture of print, gave experience a frame, providing it with a logical sequence and a sense of detachment, electronic communication is, in a sense, simultaneous and collective: it involves "all of us all at once." This is one important reason why the children of the television age differ more from their parents than their parents differed from their fathers and mothers. After all, the earlier generation shared membership in the same typographical culture, while the children in the years since the Second World War are growing up in a culture where the sensations, the conditioned reflexes, are beginning to be drastically different.

Then we must consider the impact of this high-technology society on the character of work itself. Technological change in the years since the industrial revolution has brought about a steady increase in the productivity of our economy. In the United States, for example, man-hour productivity has been growing for a considerable time at about 3 percent a year. In the past we have taken about ⅔ the gains in productivity in increased income and about ⅓ in increased free time. So the workweek a century ago averaged 66 hours in length; in 1940 about 44 hours; today about 38 hours. And if there is some stabilization in the international situation, we may expect a continued decline in working hours for the rest of the century.

At the same time, while the workweek and the work-day have diminished in length, improvements in medical technology have vastly increased life expectancy. Life expectancy for a white, male American in 1900 was 47 years; by 1940 it had become 63 years; it's over 70 today. Thus, people live longer; they enter the labor market earlier because of the extension of compulsory education; they work fewer hours in the labor market; their vacations are longer; and they retire earlier. This convergence of factors will soon produce a phenomenon hitherto unknown to history—a population that will spend more of its life at leisure than at labor.

All of this, of course, has been accompanied by changes in the kind of work people do. The United States has become a service economy; that is, it is the first society in history where more people are employed in the provision of services than in the production of tangible goods. Since 1956 white-collar workers have outnumbered blue-collar workers in the United States; today the proportion is more than 5 to 4. And, as we become a high-technology society, work grows more specialized. This produces a greater premium on the education necessary to acquire the skills for specialized work. The most startling increase among occupational groups in our society is in professional and technical employment. This was the fastest growing occupational group in the last decade; and it is expected to grow at more than twice the average rate for all fields in the next decade. Gerard Piel, the publisher of *Scientific American* has summed it up, "Work occupies fewer hours and years in the lives of everyone. What work there is grows less like work every year and the less the people work the more their production grows."

The high-technology society has thus wrought formidible changes in social organizations, in social anxiety, in human perception, and in the character of work. In particular, it has bestowed on man the astonishing gift of unprecedented amounts of free time; and it has done so at just the moment when man is intimidated and overwhelmed by the great organization, is undone by a new contagion of social anxiety, and is in a time of transition in the very way he imposes his structure on his experience. One question, therefore, is What will man in advanced industrial society do with this new infusion of free time? Or rather, perhaps, What will this infusion of free time do to man?

This, of course, leads us on to the consideration of leisure. Yet it is important, I think, to understand that free time is not necessarily leisure time. For the Greeks, leisure was, as Aristotle said, "freedom from the necessity of labor." But it was not aimlessness or idleness or frivolity or self-indulgence or lazing in the sun. Leisure, from the viewpoint of the Greeks, was educative and creative and disciplined; it was the way man grew in character and wisdom. As Emerson once said, "The soul is the color of its leisure thought."

Since humanity has had much more experience with work than with leisure, the problem of the movement of the society in the leisure direction is difficult and often distressing. What the high-technology society has thus far produced is not leisure—not what Paul Goodman has called serious activity without the pressure of necessity—but free time; and a problem for the future will be the conversion of free time into genuine leisure.

But first we must consider a related problem, and that is the probable distribution of leisure. Who will be the beneficiaries of the increase of free time in our society?

The old leisure class, the patrician elite who regarded themselves as, and to some extent were, the guardians of culture and amenity, is under the pressure of social change and seems increasingly to be committing itself to business or to good causes. Of course there remains the so-called beautiful people, the cafe-society minority; but, while they cling to their leisure, their contributions either to culture or to anything else are negligible. The old patrician groups—

who did further the interests of culture—may well have less free time under the pressure of social responsibility than they had in the past.

As for the professional and managerial groups, they will work as hard as ever: doctors, lawyers, writers, and executives are not going to look forward to the thirty-hour week in the near future. They will continue to be overworked for an indefinite time ahead until the supply of professional and managerial talent begins to catch up with the demands of an increasingly specialized society.

The people who are likely to benefit most from the new free time generated by the high technology will be the members of the lower-middle and working classes, those who for years had to labor longer fixed hours than anyone else. It is they who will have the free time and be able to use it without a sense of guilt. Even here, though, there are problems of transition. Neither the hold of the Puritan ethic nor the spur of material need has disappeared. Hence even in the lower-middle and working classes the tendency is still to use free time for overtime and second jobs, other forms of moonlighting, and so on. Yet, one also begins to detect in these sectors a new hedonism, a lessening interest in work, in the job per se as the center of life, a sense that there may be other roads to personal fulfillment, and a new and relatively guiltless relish in consumption and pleasure.

As this trend increases, society faces a tough problem. For those who are likely to have the most free time in the high-technology society are also those who, through no fault of their own, are often least trained by education and environment to use free time wisely and creatively—to convert free time into genuine leisure. The dilemma, in short, is that those whose minds and lives are least prepared for the ordeal of leisure are the ones who are going to have the most of it. A Gallup poll (April 1969) showed, for example, that 58 percent of Americans, or at least 58 percent of those responding to the poll, had never finished reading a book, excepting textbooks and the Bible, and that only 26 percent had read a book in the last month. This is why there is so much pessimism about the impact of free time in our society. Already the pessimists point out that more free time is spent watching television than in doing anything else. The increase in free time, it is feared, will place at the mercy of the mass media an ever-increasing share of the population for an ever-increasing share of their lives. The mass media, the pessimists insist, are obliged for commercial reasons to aim at the average taste. They prosper by providing a mass audience with a standardized fantasy life. They are engaged in promoting what André Malraux has called "the industrialization of dreams" and the consequence of their monopoly of taste, it is argued, will be to transform the average taste into the only taste.

The ultimate impact, it is suggested, will be to destroy people's capacity for individual emotional response, equipping a vast audience with a set of prefabricated and mass-produced reactions for every contingency of life. The pessimists agree with Max Frisch that technology is "the knack of arranging the world so that we don't have to experience it." By inundating the mass audience with substitute gratifications, the media corrupt the ability of the people to have a felt and distinctive personal experience of their own. Life is thus reduced to a series of shared clichés. People lose their sense of identity, become passive, empty, and conformist, feel safe with mediocrity, and crave

uniformity—and pull industrial civilization down with
them. Those who seek to resist mass culture will feel
either guilty because they are different, or else de-
spairing, which leads to indiscriminant rejection, se-
cession, dropping-out, and hippie revolt.

Thoreau, writing over a century ago, said in a famous passage that "the masses of men lead lives of
quiet desperation"; a stereotyped but unconscious de-
spair is concealed even under what are called the games
and amusements of mankind. Today, in what Winston
Churchill used to call "this age of clatter and buzz,
of gape and gloat," we are obliged to amend Thoreau:
Mankind now leads lives of noisy desperation. "Every
civilization," Malraux reminds us, "is threatened by
the proliferation of its fantasy life, if this fantasy life
is not oriented by values."

Given the invasion of free time by the mass media,
what does society do? Does it acquiesce in the process
of cultural demoralization? Or can it do things to promote the creative use of free time, the conversion of
free time to leisure? The question has been well posed
by Georges Friedmann, the French sociologist: "For
the great masses of mankind, it must be frankly recognized that the achievement of dignity depends on the
manner in which future societies succeed in mastering
a mass culture, which in the U.S., in Europe, and even
in the underdeveloped countries tends to smother free
time night and day." Or as Malraux put it, "Culture
is the free world's most powerful guardian against the
demons of its dreams."

This effort to master mass culture implies purpose-
ful determination on the part of a society. Now one
central instrument in such an effort, but by no means
the only instrument, is bound to be the government.

And it is this element of public policy, and its relationship to the arts, that I want to consider particularly.

The idea of a national cultural policy offends, I suppose, many American instincts. Nor, indeed, are the arts the only way in which free time can be converted into leisure. Community service, politics, voluntary association, education, sports, recreation, hobbies, travel, nature—all these offer ways of using free time to develop and enrich the human personality. Yet the arts provide, I think, the highest and the most stringent way to do this, and the arts provide the most enduring and memorable test of the quality of the civilization. Regardless of how much we may like sports, we remember Greece because of Plato, Aeschylus, and Phidias, not because of its high jumpers or javelin throwers.

The idea of a national cultural policy has significant antecedents in the American past. After all, the Declaration of Independence consecrated this nation to life, liberty, and the pursuit of happiness; and culture makes an indispensable contribution to the public happiness: that is the quality of civilization a society makes available to its members. In his first annual message in 1825, John Quincy Adams well defined the obligations of the national state: "The great object of the institution of civil government is the improvement of the condition of those who are parties to the social compact, and no government, in whatever form constituted, can accomplish the lawful ends of its institution but in proportion as it improves the conditions of those over whom it is established. Roads and canals . . . are among the most important means of improvement. But moral, political, and intellectual improvement are duties assigned . . . to social no less than to individual man. For the fulfillment of those duties, governments are invested with power, and to the attainment of the end—the progressive improvement of the condition of the governed—the exercise of delegated powers is a duty as sacred and indispensable as the usurpation of powers not granted is criminal and odious."

Not only does the idea of a national cultural policy conform to the traditions of the founding fathers, but it acquires a new point in the age in which we are moving. The electronic revolution itself, by transforming the modes of communication and calculation, will have great effects on public policy, especially by making the devolution of authority more feasible than ever before. The computer with its provisions for feedback and self-correction will, when applied to social decision, both increase the capacity of a system of self-regulation and widen the potential representation of the individual in the process. It should solve many technical problems in decentralization and participation. American political specialists have not particularly tried to figure out the impact of cybernetics on democracy. They might ponder the recent observation of Anthony Wedgwood Benn, the British Minister of Technology: "The evolution of modern management science will ultimately allow every single individual to be taken into full account in the evolution of social planning."

It should further be added that a national cultural policy is at last developing a constituency. Never before have there been so many educated Americans or so large a proportion of Americans going on to higher education. By 1970 high school enrollment in this country had increased 50 percent over that of 1960. Institutions of higher learning have about 5 million students today; they will have about 8 million in two to three years; they will have a dozen million by 1980. All this means that more people than ever before may have an interest in protecting culture against the demoralization encouraged by the mass media. The educational constituency may in time become as powerful a national interest as the farmers or other special constituencies.

In addition, there is a further reason for consideration today of the national cultural policy, and that is that the arts in America are in a state of economic crisis. The increase in public interest in the arts in recent years has not been accompanied by the broadening of the base of financial support. Costs and expectations, in spite of all the newspaper talk about the cultural explosion, have outstripped income and contributions. Between 1965 and 1968, for example, the earned income of American symphony orchestras decreased from a sum covering 62 percent to one covering 55 percent of the annual budgets, while the budgets themselves were increasing 53 percent. W. McNeil Lowry, in a Ford Foundation report on the economic crises of the arts, thus sums up the situation: "The facts are plainly that the economic crisis of the arts has lasted over-long; that private patronage, though not exhausted, is at least strained; and that except in a few cities business corporations have not begun really to help. A change in the national attitude in support of the arts is needed to help alleviate the erosion of our cultural resources and of the personal vitality of many key artistic producers."

Before beginning to consider the potentialities of public policy in this area of the United States, I would like first to consider an objection some may make: that is, the view that a national cultural policy is impossible without fundamental changes in the structure and institutions of our society. Or to put it differently, so long as so much of our society is consecrated to profit-making, it is futile to expect healthy cultural consequences.

As my argument will make clear, I would not for one moment rest the future of American culture on profit-seeking agencies. But it is conceivable that the profit motive can be curbed here, as it has been so valuably curbed in so many other fields of American life. I would go further and say that the only hope of a healthy cultural policy lies in a relatively diverse, fluid, open, pluralistic society. The more power becomes concentrated in any society, the less margin remains for the artist to pursue his own vision of truth or beauty. Those who expect that radical change in our social institutions would benefit culture are urged to contemplate the fate of the arts in the Soviet Union. There the elimination of the profit motive brought with it an elimination of all social and intellectual diversity. And it must be said on the record of history that the power motive is even more cruel and destructive than the profit motive. The Soviet Union regards artists—in Stalin's favorite term, which has not been repealed by his successors—as "engineers of the soul," and brutally punishes all those who dare create outside the party line. The fate of Sinyavsky and Daniel is only the most recent expression of what happens to art in a monolithic society.

At Amherst a few weeks before his murder in Dallas, President Kennedy said, "If art is to nourish the roots of our culture, society must set the artist free to follow his vision wherever it takes him. We must never forget that art is not a form of propaganda, it is a form of truth. . . . Artists are not engineers of the soul. The highest duty of the writer, the composer, the artist is to remain true to himself and to let the chips fall where they may." This is why any change in social structure that results in the concentration of power in the state is death to art.

For this reason the problem of government and the artist raises the most delicate questions. I would not underestimate their difficulty. For we must never forget, as the remarks of President Kennedy emphasized, that the source of art is the artist. This sounds obvious enough, yet in a day when patrons of the arts often act as if the source of art were an institution, the universities, say, or the foundation, it is no wonder that some go on to suppose that in the future the source of art may be the government itself. Universities and foundations are splendid institutions, and so too is government. But art results from the distillation of experience by a disciplined, sensitive, and passionate *individual*, possessed of an intense interior vision and capable of rendering that vision in ways that heighten and deepen the sensibility of others. It is this individual, the artist, who must always remain in the forefront of our consideration.

There is no reason to suppose that the artist will thrive under the administrations of the state. In many cases the contrary will be true. The serious artist is often a man who, while deeply committed to it, is detached from the world in which he lives; who may, indeed, in a fundamental sense, be at odds with it. Emerson, in "The American Scholar," spoke of the "state of virtual hostility" in which the scholar "seems to stand to society." And Ruskin has reminded us, "Society always has a destructive influence upon the artist: first by its sympathy with his meanest powers; secondly, by its chilly want of understanding of his greatest; and thirdly, by its vain occupation of his time and thoughts."

The artist's sacred possession is the integrity of his vision. This is something he must assert against his friends, against the institutions that wish to patronize him, against the state itself. Robert Frost at President Kennedy's inauguration spoke of

> A golden age of poetry and power
> Of which this noonday's the beginning hour.[1]

In such an alliance poetry is apt to be the fragile partner. A government which cares deeply about the artist must remember that each man kills the thing he loves.

Yet the artist can be killed by neglect as well as by love. He can be starved as well as suffocated. In industrial society, with people everywhere at the mercy of circumstances beyond individual control, the government has been obliged to enlarge its circle of concern and take new responsibility for conditions of life and labor. If the businessman, the worker, the farmer all can benefit from the efforts of government to improve the circumstances in which they work, is it not conceivable that government may also be capable of doing things that would give new freedom and opportunity to the artist? May it not lie within the power of government—the capacity to affect conditions in a way that would give the serious artist a better chance of

[1] "For John F. Kennedy His Inauguration."

self-realization? And if this were done, would not the strengthening of the artistic core of the society be even more important than strengthening the capacity, say, of people in the oil business to take unconscionable profits?

Still, when we say this, we must never forget that the source of art is the individual artist; and that the state, when it does anything which forgets this vital fact, defeats its own objective. The role of government, therefore, must be marginal. Art flourishes, we have noted, in conditions of individuality and diversity. It recoils from bureaucracy. It shuns organization. Art even flinches from organizing itself, because it knows by experience that any establishment tends to succumb to conventionality and to prefer the academic to the innovative. Creativity cannot be institutionalized, and there is no point, in culture or anywhere else, in enshrining the past at the expense of the future. "Beauty will not come at the call of a legislature," as Emerson put it. "It will come, as always, unannounced, and spring up between the feet of brave and earnest men." The notion of a cabinet ministry of the arts, I must confess, fills me with horror. I agree with John Sloan, who is said to have welcomed the idea because "then we would know where the enemy is."

But I do not think these considerations invalidate the idea of a national cultural policy. They only demand that such a policy respect the autonomy and integrity of the artist's mission, and that such a policy see its primary responsibility as preserving the highest standards of professional excellence while at the same time encouraging popular participation in, and enjoyment of, the arts. In this spirit, one can imagine a number of ways in which government can help the arts without following the Communist example of making culture a branch of politics, or transforming congressmen into cultural commissars, or setting up an elaborate bureaucracy, or establishing old and successful artists as the final arbiters of their art, or invading the privacy and integrity of the individual artist.

Let me say a word about the philosophy which should guide such programs of support. First, these are programs of *support* and not of direction or of control. We have in our country a series of acts which provide the beginnings of a national cultural policy. In 1962, President Kennedy asked August Heckscher of the 20th Century Fund to serve in the White House as a special consultant on the arts, and Heckscher's report of 1963, "The Arts and the National Government," provided the foundation for subsequent action. President Kennedy then set up an Advisory Council on the Arts by executive order, and President Johnson subsequently appointed Roger Stevens as chairman of the Council, and as his own Special Assistant on the Arts. The enactment in 1965 of the National Foundation of the Arts and Humanities Act established the two endowments; and these and other elements in the legislative framework—the Elementary and Secondary Education Act of 1965, the Public Broadcasting Act of 1967, the legislation establishing the Federal Communication Act back in the thirties—create a basis for a national cultural policy.

In addition, there are provisions in the Internal Revenue code exempting or reducing taxes on gifts or bequests to nonprofit enterprises, including those in the arts and including foundations active in support of the arts. There is also the establishment in Washington of the John F. Kennedy Center for the Performing Arts, which will provide the nation's capital with an institution designed to offer the best in opera, concert, ballet, drama, and the film, and which will work closely with other cultural institutions in elevating standards of performing arts throughout the country. Outside the national capital there are state arts councils. Today 54 of the 55 states and territories have official art agencies of some sort.

The facade is impressive. But alas, it consists more of show than of substance. The 1968 federal budget, for example, allocated only $7 million to the arts, as against $4 billion for roads and $80 billion for national security. The Ford Foundation arts program remains still the most comprehensive philanthropic activity in the arts in either the private or the public sector. Congress in 1969 sharply cut back appropriations for the endowments for the arts and humanities and for cultural exchange, though President Nixon, to his credit, has now asked Congress for $40 million for the two endowments. Very little has happened in public broadcasting; the FCC has only recently tried to exercise its powers under the Communications Act to stop the renewal of television licenses for stations which have been abusing the privileges granted them, and the industry, with the aid of Senator Pastore, is retaliating with force. Most of the state arts councils are underfunded and essentially symbolic.

Yet these activities, puny and pathetic as they are, create the foundation for a national cultural policy— one that would support and strengthen, but not regiment, the arts. These programs must be carried forward much more energetically than at present. They must be directed particularly to strengthening the professional institutions of the arts if we are to preserve our standards of excellence. Funds for artistic training in school under the Education Act are fine, but on the whole they turn out amateurs—at a time when we neglect the professional institutions of the arts, the museum, the symphony orchestra, the professional theater or opera group, the public library.

As I have said, programs involved here in the main are programs of support and not of direction or control. As the Act establishing the Foundations for the Arts and Humanities states very clearly, no government agency or official "shall exercise any direction, supervision or control over the policy determination, personnel, or curriculum, or the administration or operation of any school, or any other non-federal agency, institution, organization, or association." Within this framework, as I have said, the first effort should be to strengthen the existing centers of *professional* achievement, for it is these institutions which enforce discipline and maintain standards. Attractive as amateur symphony orchestras and local opera groups may be, the priority must surely go to the major artistic institutions, all too few of which are established on a sound financial basis. For example, for several years in the decade of the 60's the Metropolitan Opera played to over 95 percent of its capacity but emerged with a deficit of nearly a million dollars. Perhaps it could have avoided the deficit by raising the price of the tickets, but I am not sure that we want to make opera exclusively an art for the rich.

There are those, I reluctantly suppose, who would say that if a cultural institution cannot pay its own way in a free market, then it has no economic justification, and if no economic justification, no social justification. This argument implies that only those things in society which can "earn their way" in the competition of the market are worth having. A moment's reflection shows how preposterous this argument is. One can almost argue the reverse more persuasively: the most precious institutions in society—our schools, our hospitals, our clinics, our churches—are precisely those that do *not* pay their way. This does not mean that unpopularity can be taken as a sure mark of aesthetic achievement. The greatest art is great because it interprets simple and complex experience simultaneously and can thus appeal to people at many levels. But it does mean that an affluent society running the grossest national product known to history can afford to divert some of its abundance to the support of activities that enrich and benefit the nation, even if they do not meet the box-office test.

Nor need we fear that the act of subsidy per se is stultifying. The argument that political pressure is bound to win out in any relationship between government and intellectual endeavor is refuted every day by experience. Such agencies as the National Science Foundation and the National Institute of Health show that it is possible for government to make an intelligent contribution in fields almost as notorious for internal jealousy and factionalism as the arts themselves. Other countries, moreover, have subsidized the professional arts without introducing political regimentation or preventing artistic innovation. Covent Garden, for example, receives half a million pounds a year from the British Government. The French Government helps the Comédie Française. Obviously, the main support for cultural institutions in the United States will continue to come from the box office and from philanthropists. Still, when the survival of a serious opera company or symphony orchestra or art museum or public library hangs in the balance, or when it can survive only by changing prices or maintaining hours which exclude the young and the poor, then a case exists for public attention.

The danger remains of favoritism on the part of established institutions. A program of governmental assistance must guard against this by recognizing the diversity of artistic expression and by declining to surrender to cabals or cliques. This, of course, raises another danger: foolishness or philistinism on the part of the government. In the end there is no way of eliminating the factor of choice. These matters cannot be turned over to the computer. In the cultural field, as in every other area of public policy, the role of government depends ultimately on the kind of people elected to national responsibility and the kind of advice these people seek. It is conceivable that in a more tranquil future, when the harsher issues that obsess us today begin to recede from popular attention, a presidential candidate will have to state his views on cultural policy as he does on policy toward education or community development or the conservation of national resources.

The problem of government's relationship to the arts requires, in short, careful reconnaissance and ingenious planning. But it is surely nonsense to suppose that this nation cannot work out ways which will help the arts without harming the artist or infringing on his integrity. What is at stake here is not a series of cosmetic embellishments, but rather the whole quality of our civilization itself. The high-technology society will, as we have seen, generate increasing amounts of free time. This free time will largely go to those who in the past have been the least interested in the more exacting forms of cultural and artistic experience. This is not something a nation can lightly disregard. As August Heckscher has well put it, "A society is what it does in its free time." If slackness, mediocrity, and vulgarity prevail in the utilization of this free time, they will drag down the whole level of our civilization. The individual, without work in its traditional form as a stabilizer, overwhelmed by the great organization, consumed with social anxiety, disturbed by changes in the structure of perception, will be ever more dangerously adrift in the electronic age.

It seems to me that we betray our past, as well as our future, if we do not seek, as a society, to gain control over our cultural direction and destiny. A national cultural policy need threaten no legitimate interest of culture or of the arts. But it can provide a means of reinforcing the best against the worst in an unpredictable time of transition; it can offer people a chance of realizing their own best selves; it can offer our nation the chance of inscribing itself memorably on the history of the world. Let us never forget the wisdom of Ruskin: "Great nations," he said, "write their autobiographies in three manuscripts—the book of their deeds, the book of their words, and the book of their art. Not one of these books can be understood unless we read the two others; but of the three the only quite trustworthy one is the last. The acts of a nation may be triumphant by its good fortune; and its words mighty by the genius of a few of its children; but its arts only by the general gifts and common sympathies of the race." [2]

[2] From the preface to *St. Mark's Rest.*

LEO PERLIS

IMPLICATIONS FOR LABOR UNIONISM

Despite all the bitter and angry talk around us, despite all the hate and violence within us, despite all the fake prophets to the left of us and to the right of us, the simple truth is that our very survival as a united nation depends entirely on cooperation—cooperation between black and white, between labor and management, between young and old, between Christian and Jew, between one citizen and another—as Americans and as human beings.

The alternative to cooperation is confrontation, conflict, and chaos. The fact is that those who practice the politics of confrontation, the white racist and the black extremist, the economic royalist and the political revolutionist, are contributing to the sharp polarization of our country and, if we permit them, to the ultimate destruction of our system. Guns at Cornell are no less a nightmare than white sheets in Mississippi.

The democratic process itself is now being challenged by the extreme left and—paradoxically as always—by the extreme right. But the real danger will come from the extreme right. This backlash from the right will be aided by those moderates, conservative and liberal, who are beginning to react almost automatically against the massive irresponsibility of the rioting adolescents on our city streets and college campuses, who may know all the questions but who have none of the answers.

What are the questions?

Why has white America isolated and oppressed more than 20 million fellow Americans because of the color of their skin?

Why has rich America kept more than 30 million fellow Americans in relative poverty?

Why has powerful America not won the war in Vietnam?

Why has skilled America permitted its cities to deteriorate?

Why has technological America tolerated depressions, recessions, and unemployment?

Why has scientific America refused to train more physicians, more nurses, more social workers?

Why has America, the land of equality, taxed millions of its poor and permitted many of its rich to escape through legal loopholes?

Why has generous America been so callous about its first Americans, the Indians?

These are some of the questions they ask, and most of them are serious enough to require serious attention.

But while we wrangle, in comfort, over proper answers, the sick and the segregated, the oppressed and the alienated lose heart and hope and patience, and so they resign or retreat or revolt. They resign from our world into one of their own, all too often by suicide. They retreat from our reality into a fantasy of their own, of alcohol and LSD, of daydreams and nightmares. They revolt against a system which they never made and offer nothing to replace it, except anarchy.

Certainly violence is no substitute for action, and anarchy is no substitute for wisdom. And yet all this— violence and anarchy—is what passes for action and wisdom in too many young minds and at too many old meetings. It is the new morality of the incensed but inexperienced. Violence, after all, is action, and anarchy, to the young, is the better part of conventional wisdom. So they man the barricades with new revolutionary slogans and they break up meetings with old four-letter words! And while SDS (Society for the Destruction of the System) plays on, Mao, Ho, and Castro stand in the wings, applauding, cheering, conducting.

Well, what are we going to do about it?

In the face of all the problems accumulated through years of neglect, and in light of the current challenge to our system, we must first decide—in almost black and white terms—certain issues that transcend all other issues:

Do we want a united United States or do we want a divided United States?

Do we want a democratic society or do we want an authoritarian state?

Do we want a society of power and prosperity only

for some or a responsive and sensitive society designed for all?

If we want a divided, authoritarian, and repressive society, then we must encourage the extremists to extend their violent and revolutionary activities in our cities and on our campuses—until they win.

On the other hand, if we want a united, democratic, and sensitive society, we must do two things: (1) We must expose and fight the extremists, and (2) we must open the doors and windows of our institutions to all.

A united America cannot permit the existence of a white society and a black society, side by side but separate and unequal. All-black dormitories are just as bad as all-white housing. A democratic America cannot permit the accumulation of power in the hands of a few: the well-born, the well-placed, and the well-to-do. A sensitive America cannot permit poverty, discrimination, and alienation.

Let us not let any of our people go. On the contrary, let us all come in, and let us all share in the American dream, freedom to fulfill ourselves, security to sustain ourselves, and peace to perpetuate ourselves.

Where else, after all, can we go? There are, in the final analysis, only five roads open to us:

We can take the reactionary road and return to the so-called good old days of master and slave, low wages and long hours for the many, and high profits and no taxes for the few.

We can take the conservative road of retaining everything exactly as it is, all the imperfect institutions of an imperfect society.

We can take the suicidal road of resignation and sink our senses in a haze of LSD or bury ourselves in an unexplored hereafter.

We can take the revolutionary road of resistance and rebellion and destroy our system in exchange for the unknown, the untried, and, possibly, the untrue.

Or we can take the slow but sure road to reform and reconciliation, reform of our imperfect system and our imperfect institutions, and reconciliation of our individual concerns and group interests.

This last road is my road. And it is this road that I shall travel as long as it is open, because it is the only open road in an open society.

Ours is a society of hope and promise, but much collective bargaining needs to be done to fulfill the hope and the promise. I am referring to the need for extending bargaining beyond the plant gates to the total community, collective bargaining between tenants and landlords, between young and old, between students and schools, between patients and doctors, between outs and ins. And just as the strike and the picket line are an essential part of labor-management bargaining, so must protest and demonstration be an integral part of bargaining between the outs and ins—but not violence. In a democratic society physical violence is the last resort of the emotionally sick, the mentally unbalanced, and the morally bankrupt. We could do worse than isolate them for proper treatment lest they infect others. Such infection can result in an epidemic and in appeals such as this:

The streets of our country are in turmoil. The universities are full of students rebelling and rioting. Communists are seeking to destroy our country. Russia is threatening us with her might and the republic is in danger.

Yes, danger from within and from without. We need law and order. Without law and order our nation cannot survive. Elect us and we shall restore law and order. We will be respected by the nations of the world for law and order. Without law and order, our republic will fall.

This appeal was made almost forty years ago, in 1932, by Adolf Hitler.

Yes, no violence and no nonnegotiable demands. Nonnegotiable demands are pure and simple blackmail and a perversion of the democratic process. Hitler did no bargaining in Germany. Stalin did no bargaining in Russia. And neither does Mao in China or Ho in Vietnam. But can we learn from the past or are we as Santayana once said, doomed to repeat the mistakes of history?

The now generation is impatient with the past and has little faith in the future. It carries on as if it were the last generation. Living, perhaps subconsciously, in the shadow of the atom bomb and the possibility of total annihilation as a result of some irrational action, the members of the now generation challenge our assumptions, our concepts, and our conventional wisdom. Some of them damn, defy, deface, and destroy as if there were no tomorrow.

And yet, even in the shadow of the atom bomb, we must learn to live and work today as if tomorrow will come.

We all know that work has more meanings than one to both individuals and nations. In our culture, work is more than an economic necessity. Even that old fighting slogan "we don't live to work" has lost its original challenge. I suspect we have known for a long time now that work means more than money in our pockets; work can contribute also to our peace of mind and the steady beat of our heart. Our current concern with the question of compulsory retirement is a reflection of this more profound insight into the nature and meaning of work.

It is true, of course, that in different places and at different times work has had religious, political, psychological, and other implications, in addition to basic economic motivations.

It is entirely possible that the proponents of a longer workweek and a longer workday fear the softening psychological effect of a shorter workweek and a shorter workday on a people who need to know that they are engaged in the most serious struggle of our century. This is not really very different from the point of view of our puritanical forefathers who preached that idle hands do the devil's work.

My guess is that after all the sparring is done we shall settle down, possibly in our own lifetime, to both a shorter workday and a shorter workweek, seven hours a day, four days a week. And why not?

Automation is rapidly revolutionizing our industrial society. Plant after plant produces more and more goods with less and less labor. In the mining, textile, steel, auto, and other industries where the productive benefits of automation have already been felt, workers by the thousands have been laid off, many never to return again to their industries and occupations. Much of our current unemployment is due to this scientific revolution.

With a good measure of human compassion and an equally good measure of cold planning we shall surmount the current crisis and emerge from this time of uneasy transition into a period of nearly full employment, greater productivity, increased production, and a shorter workweek.

It is in the cards, I believe, that American workers will have even more leisure time in the future than they have now. Add this to the growing body of retired citizens and we have, as usual, both a boon and a problem.

It is a boon, of course, not to have to work for one's living seventy, sixty, fifty, or even forty hours a week, seven, six, or even five days a week when and if one can make a living working only five days a week at six hours a day or, even better, four days a week at seven hours a day.

Assembly-line or push-button work in modern times is no great joy in itself. There is no diversity. There is no craftsmanship. There is no opportunity for achievement. There is no chance for excellence. There is only the time clock every day and the paycheck every week.

In the long run, I see no real religious, educational, cultural, physiological, or psychological satisfactions on the modern assembly-line, push-button job.

This is not the seaman's job, and here is no challenge of the sea. This is not the logger's job, and here is no flavor of the forest. It is simply a job which must be done and for which one gets paid.

It is depressing how some men and women of wealth who work with spirit eighteen hours a day—but against a background of security with freedom of movement in a hospitable or challenging environment, at jobs they freely choose and obviously enjoy—cannot understand how a man can get sick and tired of his dull, routine job after only eight hours.

It is equally depressing when the executive who finds

his job exhilarating and challenging complains about the file clerk who does not find her job stimulating and exciting, and who is as prepared to wage the good fight for another day off as the executive is for a larger bonus.

If I belabor this point, it is simply to underline the reality that we are faced not only with more leisure time off the job, but with a duller time on the job, and that both are problems which require our attention.

The idea that something should be done with leisure time *off* the job is quite obvious. The preamble to the AFL-CIO constitution, which was adopted at the merger convention in 1955, pledges the AFL-CIO "to the attainment of security for all the people . . . and to the enjoyment of the leisure which their skills make possible."

The thought, however, that something should be done with leisure time *because of* the job is not so apparent.

There was a time when a man fulfilled himself in large part by the job he did. There are many men who still derive this sense of fulfillment from their occupations, although the complete-man concept has rapidly been gaining ground so that the job alone no longer means as much as it once did. The industrial revolution, the assembly line, mass-production—all these—helped to spell the doom of the craftsman who singly produced, with satisfaction and even with affection, the watch or the pair of shoes or the table.

It has been said that air conditioning, spaciousness, music, pastel colors all help enhance a job. I suppose they help some, but my guess is that nothing will help as much as the four-o'clock whistle—unless it is exciting and useful work off the job to relieve the boredom on the job and to give the job holder a purpose, a usefulness, a sense of fulfillment, a zing beyond the plant gates. This zing can best be achieved, perhaps, by developing an interest in continuing education, recreation, public service, and the arts.

Leisure, then, not only affords an opportunity for public service, but public service can be the best medicine for the new leisure class, the gainfully employed with time on their hands.

The untapped mental and emotional resources of the man who has no mental or emotional commitment to his job cry out for nourishment and are at the same time a potential source of mental and emotional commitment to his community and his fellow men, in short, a potential source for public service.

Now, for the first time, the average person spends more time off the job than on. He spends only eight out of every twenty-four hours, five days a week inside the office or plant. The rest is spent in his home and in his community. As an extra bonus he has two full days a week entirely free, and a number of holidays. He can either waste these free hours or make the most of them, for himself and his community.

Effective use of this free time will strengthen him and lengthen his life; the disuse or misuse of these extra hours will rob him of zest and purpose and shorten his years.

"Purpose through public service" is as good a slogan as any, but public service must be defined, on our tight little planet, as service to people everywhere, not only in Wayne Township or New York City or the State of California or even the United States of America, but in the total human community which is the world.

Here is an opportunity for our schools, organizations, and agencies to undertake programs of (a) education for citizenship, (b) social action for progress, and (c) volunteers for a better community.

These programs point in a general way to the many doors to public service which can be opened by those who are tired of watching television and for whom stamp collecting has lost its initial challenge.

The young and the old need our care.

Political freedom and civil rights need our attention.

Peace and justice demand our commitment.

A full-employment economy and economic security for all require our thinking and planning.

Our environment calls for our complete participation in programs designed to eliminate and prevent pollution of our air and water and in the conservation of our natural resources.

Here, then, is a challenge to our institutions, public and private, to help us make the most of our leisure time in public and community service, in community activities for more adequate recreational facilities and more and better parks for those who would rather spend their leisure time in these pursuits; for libraries, museums, and theaters and symphony concerts for those who would find fulfillment in the arts and culture generally.

Then, there are not many better ways for students to use their leisure time than to picket peacefully those establishments which discriminate, segregate, and isolate on the basis of race, color, or creed. This, too, is a public service.

There are not many better ways for our senior citizens to use their leisure time than to petition Congress by resolutions, delegations, and otherwise for a comprehensive, national, health insurance program. This also is a public service.

I should like to see our juvenile gangs use their leisure time to improve their neighborhoods, fight discrimination, and help patrol our schools and our parks.

We all can use our leisure time to help our students, our senior citizens, and our gangs to perform constructive functions in the public service.

From the Camp Fire Girls to the Peace Corps there are many roads to public service.

Active participation in political action—and intelligent voting, the development of community discussion groups for the purpose of evaluating legislative proposals in city hall, the state legislature, and Congress— here is one road.

Legislators and administrators of governmental agencies can help to inspire public service and at the same time extend the frontiers of democracy by encouraging citizen participation on functioning public advisory committees and commissions.

Urban development programs, low-rent public and cooperative housing, the rootlessness of our suburbs, industrial dispersion—all these require the sustained interest of citizens with time on their hands, brains in their heads, and skills at their fingertips.

Our 200,000 voluntary health and welfare agencies alone need interested and intelligent citizens for their boards, committees, programs, and fund-raising drives. Where are the citizens with leisure time who will take the time to do more than has been done in the area of community organization, planning, and action for better schools, better parks, proper jails, better roads, and the elimination of mosquitoes and graft and corruption?

And where are the shareholders who will work and attend meetings for industrial democracy, and where are the union members who will work and attend meetings for union democracy? Surely this is a good way to use leisure time, and this, too, would be a major contribution in the public interest.

A cynical newspaperman who covered the 1963 national AFL-CIO community service conference on leisure time reported that a number of delegates, even during the serious speeches and sessions on labor participation in the arts and culture, and apparently without too much difficulty, found their own way to the bar, the TV set, or whatever.

This means, of course, that there is no accounting for individual tastes in a democratic society where the options are wide and varied and where freedom of choice is accepted and defended.

Why shouldn't a worker be left to find his own way if that is what he wants? And if his own way leads him to a baseball game and a hot dog, that too is art and culture of a kind. I am told that to some a Joe DiMaggio homer was more beautiful than an Andy Warhol thing.

But, by the same token, why shouldn't those who prefer a map be given a map with suggestions on how to get there, what to look for, and how to see?

All this means is that (a) communities must provide cultural and artistic facilities and opportunities for a variety of cultural and artistic tastes; (b) workers must be given the chance to taste the community smorgasbord of cultural and artistic tidbits and to further develop their tastes; (c) unions must be concerned with the quality of life (which includes art and culture) as well as the quantity of life (which includes higher wages and shorter hours) ; (d) community art and cultural agencies, such as the symphony, the museum, the theater, the ballet, and the library, must open their doors wide to labor representation and participation; and (e) organized labor must be motivated to come in and participate actively and effectively.

This is exactly what the AFL-CIO proposes to do, in a disciplined and organized fashion, in four demonstration cities, Louisville, Minneapolis, Buffalo, and New York; and this is exactly what the AFL-CIO recommends that the labor movements in other communities do on their own.

What, for example, can a local central labor body do to help enrich the lives of its members and their families through art and culture?

Here are just a few out of fifty projects and programs we are developing:

1. Provide interested and qualified labor people for the boards and committees of community art and cultural agencies.

2. Participate in fund-raising for these agencies, possibly through the development of a Community Arts Fund similar to the local United Fund for voluntary health and welfare services.

3. Lobby for legislation and appropriations to help develop and underpin a wide variety of artistic and cultural projects, programs, and institutions.

4. Provide audiences for the performing arts through such special arrangements as union nights, block discount buying, sponsorship of special events, etc.

5. Bring the arts and culture for special events and exhibits into the union halls, plants, and neighborhoods, via mobile units if necessary.

6. Arrange for come-see tours by union leaders and members of all the community's artistic and cultural institutions similar to the union come-see tours of local health and welfare agencies.

7. Organize special lectures and exhibits on the arts and other cultural subjects beginning with those of particular interest to organized labor in union halls, libraries, museums, etc. An exhibition of protest books, posters, and paintings may be of special interest.

8. Arrange special labor tours of museums that should be kept open after working hours.

9. Encourage pioneering and innovative art forms.

These are only a few of the many projects on the drawing boards, any number of which will help to project the labor movement into an area which was once the exclusive province of the formally well-educated, the socially well-secure and the financially well-off.

A sense of moderation persuades me to suggest that the answers to the questions we pose at this conference may be found neither in one area alone, nor in another area, but in all areas of human interest.

I am not especially impressed with the either/or approach.

I take it that leisure means the constructive use of free time for personal and community enrichment. Effective use of free time will strengthen us and lengthen our lives; the disuse or misuse of free time will rob us of zest and purpose and shorten our years.

There is, of course, nothing more deadly than boredom. Boredom breeds disease among the old, crime among the young, and a sense of spiritual sterility among all of us. Boredom can be prevented and cured only by building up our inner personal resources and outer community resources.

We must, first of all, search for the answers to these questions:

What are the economic and psychological causes and consequences of moonlighting?

Is our constant hustle and bustle for things and ghosts evidence of a lack of meaning to our personal lives?

What is the relevance, if any, of this hustle and boredom to the rise in alcoholism, drug abuse, and emotional illness?

Do we really miss the good life even while we feverishly chase it? I suspect we do.

From Plato to the present, men have attempted to probe the implications of free time and the nature of leisure and its meaning to the individual and to society. The difference is that we now have more people with more free time than ever before.

It is essential, therefore, that we continue to probe deeply, not only because of the increase in free time off the job, but also because of the increase in dull time on the job.

Can we find in the constructive use of free time off the job some positive substitutes for the lack of personal satisfaction from routine assembly-line or push-button work?

I think we can.

DOROTHY MAYNOR

IMPLICATIONS FOR THE GHETTO

Let me plunge right into what I want to talk about. There is a vast segment of our United States population—the number runs into the millions—for whom the whole concept of leisure is not a positive thing at all. In fact, the word leisure is hardly used in the ghetto. In communities such as Harlem, or Bedford-Stuyvesant, or Watts, or Newark's Central Ward—these are areas where Negroes, Puerto Ricans, and other marginal Americans live—in these places we speak of "unemployment," or "idleness," or "shiftlessness." The word "vagrant" is beyond the reach of most of us; but a fellow without a job is sized up as a "drifter," a "beggar," a "hustler," a "tramp." In our quarter of the American culture, if a person

is not holding down a job requiring at least forty hours
of his week on a regular schedule, he is suspected of
being a trickster or maybe a thief. Actors were once
classed with the most disreputable element of the population because they were without steady jobs.

Of course it is a deeply ingrained feature of our
Puritan inheritance to glorify work. Idleness and mis-
chief were generally paired. I don't know that anybody
takes Kipling seriously today; in fact, I rarely hear
him mentioned anymore. But he said

> If you can fill the unforgiving minute
>   With sixty seconds' worth of distance run—
> Yours is the Earth and everything that's in it,
>   And—which is more—you'll be a Man, my son!

When he said that, he was speaking the language of
Western man; he was laying bare the secret of the
English-speaking world. Or so it was imagined.

Hence, idleness came to be viewed as a form of im-
morality. The summons to toil was sounded at sunup.
The man who ignored that call had no standing with
his neighbors. John Wesley is remembered more for
his toil than for his virtue. Generations of Methodists
were encouraged to remember him as one who was out
of bed every morning at four o'clock. As little as I
know about such things in detail, my understanding is
that one reason the early labor leaders in this country
and in England had such a hard time was that they
were opposed by men who professed to believe—and
doubtless did believe—that in asking for more free
time for the masses they were asking for something
that was somehow contrary to the will of God.

Children began working, in field or factory, at a
very early age. Old men worked right up to almost the

day of their funeral. To be on earth was to be under obligation to work.

This has very largely changed within the lifetime of some of us. The reasons for this change do not need to be dealt with at great length here. It is no longer regarded as shameful that a man in such a social order as ours is not glued to the drudgery of menial tasks, some of them degrading to mind and spirit, from sunrise to sunset, practically from the cradle to the grave. To be free from the demands of work and duty, to have at one's disposal unoccupied time, it is now freely conceded, is a prerogative of the modern free society.

In my day-in, day-out experience I am working with a segment of the American population that has never yet been encouraged to think *positively* about this whole matter of leisure time. The reasons for this are complex beyond telling, and are rooted in some of the most tragic phases of our history as a nation.

It is scarcely reasonable to expect easy and immediate solutions to problems that have been allowed to fester for generations. These problems are as old as the perfectly understandable attitude of the slave whose sweat is drawn from him by means of a lash. They are as new as his exclusion from the advantages of membership in a labor union. They may be seen in the fact that by the tens of thousands the Negro is being turned off farms in the South, where his limited skills are no longer needed, leaving him no place to go but into the already overcrowded slums of the cities.

I am merely saying—and I am sorry if this makes you think of Banquo's ghost—I am merely saying that if we are to discuss with any seriousness the question of the relation of American technology to human values and leisure, we are not going to get very far into this subject before we run into the matter of what bearing this has on the lives and fortunes of some twenty-odd million black people in our midst. For here is involved the fate of our cities, the health of our economy, our tax structure, and our posture before the world as a Democratic society.

I want to try and bring home to you the scope and nature of this problem, a problem into which I immersed myself some five years ago, at the time of my retirement from the concert stage.

The Harlem School of the Arts, now just out of its infancy, and moving on, as one would hope, to bigger things as an independent agency with prospects for increasing usefulness, has up to the present time had to lean heavily upon the aid and the resources of such friends as we were able to find. I know there will never be a time when we shall not need friends, those we have already, and whatever others we may be able to cultivate. But in these first five years we have been like a homeless beggar, with no money, no building, no preconceived notion, actually, of what we needed to do.

The church of which my husband is the pastor gave the school quarters and took care of the cost of heating, along with some other basic needs. As for the rest, we were on our own.

It is pleasant to be able to report that there were from the outset, as I soon found out, a great many people outside the Harlem community, white people, that is, who were willing to help with their means and their counsel. We got together a board of deeply concerned men and women, some white, some black, who were willing to give of their time and understanding. These friends made other friends; so that now we have in our files some two thousand men and women of various backgrounds and circumstances. All these, in one manner or another, some in large ways, others through small tokens, but all in some fashion, have shown their interest in this modest effort in behalf of the six hundred boys and girls registered with the school, whose quest for some measure of enrichment to their lives, such as we are trying to give them.

There is nothing new about this, you may say. And I would agree with that judgment up to a point. It is a fact that educational projects for the disadvantaged have had sympathetic response from the more fortunate all through our history as a nation. I think I see in what we are getting in Harlem—and I would hope that this could be duplicated in other such communities, North and South—something rather different from what we have all too often had in the past, when a sop was sometimes thrown in the direction of the Negro, maybe out of pity or contempt.

I honestly believe there are a number of people— not enough, but, one would hope, a growing number— who have at last come to feel that something really fundamental and creative has to be done about the lack of equality that has disfigured so much of America's history. There is a lot in the news every day that is grave enough to keep us awake far into the night. We hear words such as "confrontation" and "polarization," expressions that seem to say that the lines are drawn and we are about to witness more bloodshed than anything the past can show.

We well may be on guard against the worst that could happen. If Harvard can make itself ridiculous, what can be expected from the rest of us? Let us not be bemused or misled by wishful thinking. There is a lot of tough talk in the air. We are saying terrible things to each other. Anybody with a nose for disaster ahead must find this an age of great anxiety. There are those who insist that we are headed into the teeth of a storm, and that we need now to be occupied with trying to brace ourselves against impending doom. I think the worst can be avoided. I believe there are within both black and white America resources of understanding and practical common sense (notice that I am not running the risk of your ridicule by speaking of "good will"). I do believe we are hardheaded enough, sensible enough, pragmatic enough to do what we have to do to see to it that we keep this ship afloat and seaworthy. Whatever is necessary will be done, done perhaps not in the name of righteousness or to bring in the kingdom of God—these are concepts from which in recent years we have tended to shy away—but in the name of decency and common sense and self-preservation.

A great many people, not all of them in New York City, but some even across the continent, have heard of our effort to extend the hand of helpfulness to the children in Harlem and have shown their interest in various ways. This encourages me to hope that more and more of us are serious about the responsibilities of citizenship in a Democratic society.

What effect is our effort having upon the children and their parents? I remind you, of course, that we have had only a brief history. We have no miracles to report. We have had our share of disappointments and failures, but these have been the exceptions. By and large, speaking for the boys and girls, and for their parents as well, the investment we are making in the life of our community is abundantly justified. I could relate particular instances where it is clearly demonstrated that what we are putting in is yielding a very satisfactory return.

Given the incentives, these children can learn as well as anybody. And their parents are pathetically eager to see their boys and girls come into something larger and finer than they themselves have ever been blessed to know.

Of course, we are in a wasteland there. And it is so with the poor and the disadvantaged in all our urban places. But we do not need to confine it to the cities. Were you ever on some of our Indian reservations? Did you ever visit Appalachia? Do you know the worst of rural Mississippi or South Carolina? We call ourselves a rich nation. I suppose it is possible for an average citizen of this favored land to spend his whole life in ways that shut out or at least disguise the more unattractive side of this nation's life. If you have an expense account and a pocket full of credit cards and half an acre in the suburbs, you still have a lot to worry about. But you may find it hard to believe that there is starvation a whole lot closer to you than Vietnam; that there are tens of thousands of people right here in this land of abundance, not all of them black, not all of them lazy, who go to bed hungry every night. But that, so recent investigation has brought to light, is the humbling fact.

The reason I am a little more hopeful of the future than some of my friends appear to be is because I believe we are a little more open-minded about certain aspects of the human condition than we used to be. I think I see some signs of a growing maturity on the part of the American people. I don't know what you thought of the results of the 1968 election, but the American people were not stampeded by extremists.

George Wallace tried, and he simply was not able to bring it off, although, in the eyes of a lot of people—and I admit to having been one of them—he had considerable provocation working for him. But common sense prevailed.

When we speak of those who ought to be getting more of what every American has come to regard as his birthright, we know there are contrary opinions—and a certain amount of complaining—about the lazy being pampered. But we are less likely now than back in the days of the late Sen. Joseph McCarthy to label as a Communist serving the interests of Russia anybody who dares complain about things as they are.

I think we are much more ready nowadays to concede that men are frequently as they are because of conditions beyond their powers to correct, that the human spirit comes to life only when it is not stunted by surrounding circumstances.

I imagine many of you know the Nile Valley. You may have been startled upon first seeing the difference between the area that is irrigated and that not artificially supplied with moisture. Where the water is brought in, the land is fertile and productive. But right alongside this strip made green and lush by irrigation, there is nothing but barren sand, not a blade of grass or a square foot of shade.

Efforts such as that which I am undertaking in Harlem are irrigation projects, in a sense. They sometimes seem feeble and inadequate, and I would be less than candid if I denied that in these five years I have ever had a moment of doubt about the fruitfulness of such a venture. The effort is so small, so incidental to the total need. I just wish the heavens would open, so that this meager irrigation thing could be closed down.

But such vain wishing will not change a thing. Meanwhile, I have the satisfaction of knowing that, as a result of the combined efforts of many people, at least there is some grass growing where before we started there was nothing but barren soil.

I would conclude with the hope that means may be devised whereby one of the nation's great assets may be tapped in the effort to get done the many urgent human tasks facing us today. That asset, comparable to the greatest of our resources, lies hidden in the minds and hearts of tens of millions of concerned Americans who, their careers successfully concluded, are simply waiting to be called, in the words of John Milton, "to fresh woods, and pastures new."

ROBERT M. HUTCHINS

IMPLICATIONS FOR EDUCATION

The six propositions that I should like to advance are as follows:
1. Science and technology are here to stay.
2. We are entering a post-industrial age.
3. Science will create new wealth, and machines will do the work.
4. Values and institutions associated with industrialism will disappear.
5. A new era of leisure can dawn; the society could become a learning society.
6. All this is contingent on the control of the applications of science on a worldwide basis.

The first proposition is that science and technology are here to stay. This is a more sensational proposition than it would appear. The Greeks developed a theoretical foundation of science but did not press their applications; the Chinese developed the theoretical foundations of science, but did not press their application. They seem to have thought that in some way these applications were beneath them, or even immoral. The Greeks certainly believed that they were dangerous. The Arabs, after a brilliant beginning, turned their backs on both science and technology. There is, therefore, no prior reason for believing that because a scientific and technological civilization exists it will continue to be scientific and technological.

We have seen and felt the benefits of science and its applications; we have grown used to the horror and anxiety that accompany these developments. And, in any event, science and technology are now regarded as essential to the national power and prosperity. It seems inconceivable that we, like the Arabs, shall turn our backs on either one.

This brings us to proposition two. We are entering a post-industrial age. This is an age in which the material wants of mankind can be satisfied with very little human labor. Everything is radically affected by this change. This is, I think, the fundamental change in the human condition, and the present worldwide disorders intimate a consciousness of this change throughout the world. The change will affect the attitudes of all people toward work, business, and competition. The change will raise the question What comes after affluence? It will raise the question What do we do with power? The aims and expectations of the industrial past are thus seen as inadequate, and this is because of proposition three.

Science will create new wealth and machines will

do the work. As long ago as the end of the last century Hilaire Belloc said, "The path of life, men said, is hard and rough merely because we do not know enough; when science has discovered something more, we shall be happier than we were before." This dream, which has seemed ridiculous, may now become reality. This is a change from limited to unlimited resources. The necessity of fighting for resources has disappeared. It is further evident that if we do fight over the resources that are available we won't survive. The consequences of this change also are of the most far-reaching importance. For example, the nation-state has been the sponsor of industrialism, the engine of the rapacity and greed of the industrial society. The nation-state is essentially a warfare state. Industrialism has depended on the manipulation of the masses and procuring their dedication to the industrial state.

So we come to proposition four, which is that values and institutions associated with industrialism will disappear. I think we have come to see at last that technology is not what it was supposed to be one hundred years ago, a substitute for justice. It is not the opiate of the people. As to war, we can now see that it is impossible as well as unnecessary. Patriotism takes on an entirely different content. You can no longer speak of America first. You have to speak of humanity first. Science as a source of power and domination ceases to be interesting. Sir Karl Popper, in the March 1969 issue of *Encounter*, made it stronger than that. He said, "As far as science is concerned, there is no doubt in my mind that to look upon it as a means of increasing one's power is a sin against the Holy Ghost."

Work ceases to be good in itself. And for those in my generation this is the greatest change of all. In my

boyhood, *Horatio Alger* held the place now occupied
by comic books. His heroes were not very bright. I
can't remember that any of them had any more education than the law required. But they were industrious.
And those of us who followed Horatio Alger's path
knew that the way to success was to be industrious.
Now machines are going to do the work, and a new
kind of creativity is open to the individual.

Education has been the processing of the young for
the industrial society. The educational institutions of
this country have been engaged in training for that
purpose and in holding it to the young as a means to
wealth and power. The multiversity is the characteristic educational phenomenon of our age. The multiversity is a collection of technical schools increasing
the wealth and power of the industrial state. It is a
response to the parallelogram of forces. The multiversity, dedicated to the purposes of the industrial state,
will, as the industrial society disappears, have to go
with it or remain as a memorial of an outworn age.
The post-industrial society is characterized, of course,
by a massive increase in the free time that is available
to the population. One of the most interesting questions is, What kind of work is the population supposed
to do in a post-industrial society? I take as a symbol of
that society the Bahlsen bakeries in Hanover, Germany. The manager-director said, "Here the skill of
the baker dies." And there are, in fact, no bakers in the
bakery. There are a few chemists, physicists, and other
magicians at the top. The labor force consists of illit-
erate Spanish women who ride bicycles back and forth
in front of the ovens waiting for a red light to go on
or a whistle to sound. Then they notify the repairman

to come and fix up the ovens. Since they can't speak German, they do this by pressing a button.

The question is, Is this the labor force that will be required in the post-industrial age? If so, what do we mean when we say that we have to have an entirely new level of advanced technical training for the labor force in the new post-industrial society? As far as I can see, the only training that is required for the Spanish women to whom I have referred is training to ride a bicycle.

We come then to proposition five. The society could be a learning society. I must say at once that it doesn't have to be. We do not have to realize the human possibilities of the post-industrial age. We do not have to have a learning society. What we could have instead is 1984 or Brave New World. There are some indications in my part of the country that there are important people who would like that situation. The aim of society could be the fullest human development of the individual and the community. And that community could be a world community. Education could be seen as a lifelong process. In fact, education might be, in the post-industrial age, largely noninstitutional, or it might be intermittently institutional. The computer and other devices could make every home a learning unit. One of my former associates at the University of Chicago is now a dean at the University of California, and he has seen a totally new vision of the university. He has a computer on every floor of every dormitory of the University of California at Irvine. And he looks forward to the rapid expansion of these computers as their price declines and his resources increase. I said to him, "If you are prepared to put the computer on every floor of every dormitory, and, ultimately, as the price declines and you get richer, to put one in every room of every dormitory, why go to the trouble of maintaining this plant? Why not put the computer in every home?" "Oh," he said, "that's what I intend to do. The campus at Irvine will be obsolete in ten years." Nobody knows what the limits of the computer and other devices are. Nor does anybody know the rapidity with which my friend can carry out his ambitions for Irvine. In this situation, teachers might function, as doctors and visiting nurses do today. It is, after all, ridiculous to set aside certain years of life, certain places, certain rooms or classrooms or lecture rooms for a process which by definition in a learning society becomes the principal preoccupation of all the members of all the society all the time. Educational institutions would remain, I hope, because they would provide places for the interaction of minds.

I assume you all agree, since you have all attended educational institutions, that at the present time the educational institutions of America are counter-educational. The reason is that education in any definition that makes any sense is inconceivable unless it provides for and promotes the interaction of minds. And I think you will agree that the educational system of the United States and of most other countries, at all levels, does not make such interaction possible; in fact, it prevents such interaction that might otherwise occur. But if institutions could be established for the purpose of promoting the interaction of minds instead of processing people for the industrial society, instead of trying to give them the vocational certificates that they require in order to get admitted to one occupation or another, like mortuary science, then it would be possible for the interaction of minds to take place in these institutions.

If you look at England today, you see a dim sort of forecast or prediction of what could happen. Here is a country which is, from an economic point of view, flat on its back. The Labor Party is certainly flat on its back. Everything has been cut except one thing: the proposal for the Open University. The proposal could have been abandoned with great ease because nobody but Jenny Lee and Harold Wilson were behind it when it was proposed several years ago. The planning committee has just reported; the vice-chancellor, the chancellor, the deans have been appointed. The structure has been approved. And within one more year, unless some greater catastrophe overtakes the country, the Open University will open. The headline in the last British educational magazine that I saw was "The Wide-Open University Opens"; it is open, wide open to everybody all the time and by every technological device (radio, television, computers) that the government can summon to its assistance. And it is expected that 150,000 students will enter in the first year. You get some idea of how significant this is when you realize that the total admission to universities in England today is 50,000 a year. This is an institution, then, of a noninstitutional sort, an institution of a sort with which we are not familiar, an institution in which the object is learning—in every place, in every way, by every method that can be devised, without regard to traditional institutions or the counter-educational practices that they have developed.

I must confess that I do not believe in the future of the fun society, nor do I think the future of the work society, in the sense in which we have used the word "work," is at all bright. If the fun society is contrary to human nature, and if the work society is contrary to the necessities of mankind in the post-industrial age, the only society that is left, I think, is the human society. If you analyze what it means to be human, you will be content with the definition of the society as a learning society. When manpower is unnecessary, manhood has to be the aim. This view of what could happen in the post-industrial society is not Utopian, because it does not depend very much on the intelligence and character of the residents of this planet, if they can survive. Industrialism seems to me to have set in motion irreversible tendencies that will lead to its extinction; it has dug its own grave.

The intelligence and character, therefore, that are required have to be directed to survival, as suggested by proposition six. Proposition six is: This future, or this bright future, is contingent on the control of the applications of science on a world scale. At present, we must feel that if our enemies don't get us, our neighbors will. We can expect to be suffocated, run over, trampled to death, or incinerated at any moment. Nobody who has tried to breathe in Los Angeles or swim off Santa Barbara—I've had too much sense to do that—or drive on any freeway, as I have to do every day, can be ignorant of the dangers that are just around the corner, perhaps even closer. An Australian judge has said that if the tremendously successful economic development of Australia continues at the present rate, in a very few years Australia would resemble a disused quarry surrounded by a vast and malodorous oil slick. Now I am not one of Kenneth Boulding's pessimists who see every technical advance as increased opportunity to do evil, but anybody who has lived as long as I must deeply believe in original sin. We do not know how to control ourselves. I do not blame the scientists or the technologists. We do not know how to control ourselves. The control of the applications of science has never been accomplished; we do whatever we can do. We can't always tell what the consequences of what we're making will be. The foundation of ethics is foresight, and, if you can't see the future, how can you tell if it's right or wrong? The most innocent effort may turn out to have the most dreadful effects. At the present time, we do what we can, whether we need to or not, in order to show we can do it—usually to keep ahead of the Russians.

The history of the operations on the continental shelf since 1959 is highly suggestive: The agreement reached in 1959 restricted operations on the continental shelf to 250 meters, or the limits of technical exploitation. At that time, there was no technical exploitation anywhere near 250 meters. Since that time, technology has advanced so rapidly that you hear of technological development going on at a thousand meters or any figure you like. The riches of the continental shelves seem to be greater than was assumed in 1959. Therefore, the first thing that happens is that technology outruns the political devices that were designed to control it. In a meeting of the Center for the Study of Democratic Institutions in Santa Barbara, the ambassador to the U.N. from Argentina said that the continental shelf of Argentina, according to his present calculations, was one million, five hundred thousand square kilometers, and, he added, very rich. It was perfectly plain from what the ambassadors at the meeting said (a dozen of them were present) that every nation was going to extend the boundaries of its continental shelf just as rapidly as it appeared there was anything worthwhile at the bottom. It was also evident that technology was advancing at such a rate that if there was anything worthwhile at the bottom, they could get it out. The redefinition of the continental shelf in such a way as to eliminate the ocean would be to eliminate the last common property of mankind, which is the deep seas. Not only would the extension of the continental shelf in such a way as to wipe out the seas deprive mankind of its last common property, it might also transform this last common property into the scene of Armageddon because of the appropriations of this property that are now proceeding, about which the United States is asserting itself in the most dogmatic way.

We are now confronted with the most terrifying concentration of instruments of indoctrination and annihilation in history. So much is this the case that Eugene Rabinowitch, in an article in the *Bulletin of the Atomic Scientists,* demanded an international revolution against the worldwide establishment of sovereign nations and called on the scientists and youth to lead this revolution. I hope those of you who are either or both will respond.

Is the future of mankind really in the hands of mankind to control?

The whole basis of the doctrine of original sin is that man is capable of anything. The fact that he is presented with the new situations, new facts, new developments is nothing but a challenge to his extraordinary inventiveness. The problem, it seems to me, is that we can't outgrow our old habits or old attitudes. We are too busy and apathetic

to notice what's going on in the world. If we understand
what's going on in the world, then we can do something
about it. Who would have said, a few years ago, that the
students and professors of the United States could over-
throw the government of the United States? They did, as
we learned from a broadcast on March 31, 1968. So the best
motto is the one given to me by Jacques Maritain, the
French philosopher, who attributed it to William the Silent.
He said, "It is not necessary to hope in order to undertake,
nor to succeed in order to persevere."

*Do you see any improvement in the American university
during the past fifteen years?*

Certainly nothing has happened since 1955 that would
lead one to suppose that the university situation has im-
proved. It's fair to say that an era ended at the end of the
last war. During the war the scientists had shown what they
could do, they could blow up the world. The university was
formerly a refuge for the rich, with scholarship boys to
make it look democratic, on the edge of society. Nobody
cared about it, nobody paid any attention to it, industry
didn't support it, government didn't support it. This institution, long neglected, long insignificant, reasonably honest,
moved from the periphery of society to the center. President Kennedy, President Johnson—every president—an-
nounced that the power and wealth of the United States
depended on the development of this central institution.
At the same time, tremendous numbers of students of
very mixed backgrounds moved into the institution. The
university was no good before the last war. The difference
is that nobody cared. The university wasn't important. It
was a place that took the sons of the rich and rendered them
harmless to society. Now you have all the difficulties that
existed before the last war—specialism, departmentalism,
extracurriculum activities, intercollegiate football—plus the
enormous complications of government contracts, research
development industrial development, and a number of heter-
ogeneous new kinds of students. This institution, now central, has all its old problems and all these new problems.
Because it is central, with everybody looking at it, it is, I

should say, definitely in a more precarious or desperate situation now than it was in 1955.

What is your view of leisure?

Leisure is, in my view, a normative word. The neutral word is free time, or vacant time. What we are getting with technology as it has advanced is vacant time. The job is to turn it into leisure. We have not been successful in the transformation in my lifetime. When I was a boy, people were working in the steel mills of Gary 60 hours a week. The hours of labor have been reduced by ⅓ throughout the country and we cannot say the cultural level of the country has been raised by 33⅓ percent. What has happened is that the vacant time has remained vacant and, in this, television has helped a great deal. Now, the transformation of free time, vacant time, into leisure is effected by the individual's using the vacant time as a person, a thinking human being, and the development of the community as a political, social, economic, cultural unit. The purpose of learning in the post-industrial era is to become a human being again. It is to learn for the purpose of understanding—understanding yourself, your society, the world, the methods of judging yourself, society, and the world.

What is the role of the elementary and secondary school, and do you fit it into a post-industrial society or leisure society?

Of course you can't, the schools are counter-education. The elementary and secondary schools have got to be changed so that they become educational. For example, the fight that has gone on in the Supreme Court of the United States with Mr. Justice Fortas writing the majority . . . and Mr. Justice Black vigorously dissenting: . . . Mr. Justice Fortas has said the object of education is not to establish a closed-circuit system where students repeat and memorize and regurgitate the current slogans of ideas as is suitable at their age. Mr. Justice Black, in dissenting opinions, has said that children are to be seen and not heard, that they're to keep their noses in their books and shut up, and that the rest of the Justices of the Supreme Court ought to know what he knows; namely, that the young people of this country are already on the rampage and ought to be suppressed. I think that Mr. Justice Fortas was roughly on the beam, and that this is the kind of education that he was talking about.

Should young people "cop out" of the education system?
Aristotle said that all men by nature desire to know. This is the opening sentence of his *Metaphysics*. And I think it is a fundamental educational proposition. Many of the dropouts from our present society seem to be taking the position that there is nothing to learn, and that you can be satisfactorily human with a combination of vegetation and drugs. I don't believe it will work. I don't think it's possible. I don't even think that anti-culture is possible. My friend Phillip Rieff, a sociologist at the University of Pennsylvania, has developed a theory of total, absolute revolution, in which the sole object of the individual and the group is disruption, continuous disruption. I don't believe it's possible. Even those who are engaging in disruption have some scale of values, some set of standards that they hope the community will adopt.

What would you propose as a course of action in the light of your criticisms?
I think you have to do something about the university. The object of the university negatively stated is to tame the excesses of the specialists, each of whom believes that his own subject is the most important thing in the world. I think you have to do something about liberal education. I think you have to do something about the restoration of philosophy, which has been degraded into linguistic analysis. Its principal courses are historical; you can do a better job reading the Encyclopaedia Britannica. If everybody had a basic liberal education, we would at least be aware of many problems that many of us are not aware of at all now because they were never brought to our attention. They don't arise in the course of our specialization.

I think you have to do something about politics. We have to get over the idea that politics is power. Politics is the science of the common good; and the degradation of political science in this country prevents us from ever considering the problem that you are talking about, namely, how do we understand and achieve the common good? Finally, I think you have to do something about this issue that I mentioned before, the issue of the professions. All occupations want to get social standing. So what they do is, they get a law passed providing that schools shall be established at the state university in a certain occupation, say anatomical engineering, then provide that nobody shall practice anatomical engineering in the state unless he has a degree from the school. Social standing is then achieved, competition is limited, but there is nothing to teach, there is nothing to learn, and the university is simply more confused than ever.

PHILLIP BOSSERMAN

IMPLICATIONS FOR YOUTH

Youth signal the
emergence of a new type of society. Traditional agriculture cultures moved directly from childhood to
adulthood. There were great amounts of work to be
done. Life expectancy was short. The rites of passage
ushered children into adult roles with elaborate ceremony. There still exists, in a very truncated form, the
"bush school" of certain West African cultures. The
Poro and Sande secret societies into which children
are initiated and thereby "become adults" had a special
period of instruction for children of a certain age that
often lasted four years. When children entered the
school it was cause of great ceremony. They anticipated this event with their families. There was a sense

131

of awe and fear, for the children were to become privy to the innermost secrets of the tribe; they were being ushered into the inner sanctum of the tribe's whole spiritual life. Here were the most sacred symbols which gave cohesion and purpose to the group, literally, the foundation of the society. No wonder this was holy ground and so brought with it anticipation, fear, excitement. When it was time for the bush school, the children with their parents followed the drumbeats of the "devil dancer" to the entrance of the school. They were thrown over the fence. They concealed a bag in their clothes with animal blood in it, which burst upon hitting the stakes of the fence marking the entrance. The spilled blood symbolized the death of the child. When he would emerge again from the sacred grounds of the school, he would do so as an adult. At that time he would be given a small carved mask that represented his adult role. He would carry this with him as a reminder of who he now was.[1]

The bush school is rapidly disappearing. A new ritual is appearing as industrialization and urbanization come to West Africa. It is the formal, western type school which is training children in the skills necessary for industrialization. The new African nations are experiencing "modernity" and with it the emergence of adolescence. The new school marks a period of preparation for productive work. It must now be both the bush school and the industrial-skill-producing institution. In the old bush school the child not only learned the sacred symbols and ways of his people, but also learned

[1] The information about the bush school comes from field notes of the author during work and research in West Africa, 1963-1969, and from Cecily Delafield, "The Bush School" (unpublished), 1968.

the practical skills connected to his role as a male or female in an agricultural society. Life and learning were closely linked together. Knowledge came from practical experience.

The new public or private, western type school tends to separate learning from practice. The child is taken from his family life or village. He is put in school for long hours each day, for several months a year, and for a number of years more than the bush school encompassed. The adjustments are difficult. Parents complain that children no longer have respect for the old ways. They are "frisky," they do not obey their parents or their teachers.

The point of this brief description is that adolescence is a product of industrialism. Such a period in a person's life is necessary in an industrial society before he enters adulthood. What we now witness is a new period of development between childhood and adulthood, a period called *youth*. Just as adolescence in its way was part and parcel of a new type of society, so youth augurs the emergence of a new kind of society. The period of youth is a disengagement from societal obligations. It follows adolescence and represents a time of additional preparation for adulthood. The type of training may be in the form of service, travel, or involvement in new experiences. The requirements of the new society are different, just as they were for the industrial type.

Disengagement really means leisure time. Significantly, there is the free time available to link life and learning. This is particularly true of youth in colleges and universities. However, the structures of those institutions do not allow such linkage. Sensitive students know this. How and why?

Traditionally the university has always been a place where freedom of thought exists and the total society could come under the critical eye of curious intellects. In many respects, that critical, Socratic role still persists. However, there is another aspect to schools which must be underlined. This grew out of the separation that occurred at the time of industrialism. Learning became divorced from one's total life. Though comprehensive courses and schools were supposedly developed, their comprehensiveness was related to keeping students happy and helping them be potentially good producers and consumers. Ivan Illich analyzes this characteristic in the following way:

The institutionalized values school instills are quantified ones. School initiates young people into a world where everything can be measured, including their imaginations, and, indeed, man himself.

But personal growth is not a measurable entity. It is growth in disciplined dissidence, which cannot be measured against any rod, or any curriculum, now compared to someone else's achievement. In such learning one can emulate others only in imaginative endeavor, and follow in their footsteps rather than mimic their gait. The learning I prize is immeasurable re-creation.

School pretends to break learning up into subject "matters" to build into the pupil a curriculum made of these prefabricated blocks, and to gauge the result on an international scale. Men and women who submit to the standard of others for the measure of their own personal growth soon apply the same ruler to themselves. They no longer have to be put in their place, but put themselves into the niche which they have been taught to seek, and, in the very process, put their fellows into their places, too, until everybody and everything fits.

Men and women who have been schooled down to size let unmeasured experience slip out of their hands. To them,

what cannot be measured becomes secondary, threatening.
They do not have to be robbed of their creativity. Under
instruction, they have unlearned to "do" their thing or "be"
themselves, and value only what has been made or could
be made.

Once men and women have the idea schooled into them
that values can be produced and measured, they tend to
accept all kinds of rankings. There is a scale for the development of nations, another for the intelligence of babies,
and even progress toward peace can be calculated according
to body count. In a schooled world, the road to happiness
is paved with a consumer's index.[2]

Industrialism brought with it the values of frontier
individualism applied to the business world. Competi-
tiveness became a natural attribute of man. Profit developed as the motivating force of industrial growth.
Each man was on his own to make it or to fail. The
fittest should and would survive. Hard work, frugality,
personal morality, and punctuality became the means to
profit. It is easy to see how everything became quantified. Progress was economic growth. Bigness and the
"body count" became the criteria of success. Again
Illich brings home this point:

The Vietnam war fits the logic of the moment. Its success has been measured by the numbers of persons effectively treated by cheap bullets delivered at immense cost,
and this brutal calculus is unashamedly called "body count."
Just as business is business, the never ending accumulation
of money, so war is killing—the never ending accumulation
of dead bodies. In like manner, education is schooling, and
this open-ended process is counted in pupil-hours. The various processes are irreversible and self-justifying. By economic standards, the country gets richer and richer. By
death-accounting standards, the nation goes on winning its

[2] "Schooling: The Ritual of Progress," *The New York Review of Books*, December 3, 1970, p. 22. Reprinted with permission from *The New York Review of Books*. Copyright © 1970.

war forever. And by school standards, the population becomes increasingly educated.[3]

The category of *youth* is the end result of schooling.
Youth encompass two contradictory, paradoxical responses. Youth are at once the products of the system,
schooled to be both consumers and producers of more
consumers. They are also inheritors—as vague and di-
luted as this tradition has become—of social criticism
and intellectual curiosity. This critical edge, this Socratic touch has left its mark. The uprisings of 1968
on the campuses throughout the modernized world
speak to this. Youth know that growth—real personal
and social growth—cannot be measured. If it is quantified, it is insatiable. One always feels something is
missing. Kenneth Keniston recorded this revealing
statement of a young respondent: "You don't know
how bad it really is until you've seen Scarsdale." A
quality of life is missing. This notion of quality cuts
two ways. First, the human psyche suffers. Second, the
insatiable hunger of an economic system based on
constant growth makes it consume resources at all
costs. This has resulted in the environmental crisis
that is developing into a worldwide phenomenon. Quantitative growth is destroying the very source of the
good life. Hence, both spiritually and physically, unlimited quantitative increase makes impossible a quality to life essential for psychical and physical well-
being.

A Dual Revolution

Youth are disengaged from involvement in the adult
world. This disengagement is by choice and the result

[3] *Ibid.*

of technological innovation. Hence, the sources of the dual revolution. Taking the last first, technology, especially in the form of cybernation has made possible the time for this disengagement. Greater preparation is essential to handle this new technology. The extension of schooling is the organizational response to this requirement. Just as adolescence was created by the skill needs of industrialism, so youth is the creation of the new technology. Only this time around, these skills are really not needed in the proportion as before. The machine can simply do more. Cybernation extends the effectiveness not only of the body, but of the central nervous system. It functions increasingly for the human brain. Hence, man's power over nature has grown dramatically. Likewise, his control over society has enlarged. Bureaucracy, aided and abetted by cybernation, makes this possible. This brings with it a heightened sense of manipulation and powerlessness.

Cybernation makes it possible to wring more from the environment. Nature is brought under increasing control. Cybernetics is a technique, a tool. Coupled as it is with the productive system, there are few moral guidelines for its use. The technique is subservient to the aims of production and consumptive growth. Linked with the organization of the new technology—bureaucracy—the possibility of control of human beings is enormous. Man increasingly senses being repressed, choked off, manipulated. Paradoxically the new technology *is* more efficient, more rational, able to deliver more—to up the "body count," so to speak. Abundance and free time are direct products.

The canned quality of "joy," for example, Christmas joy, results from this new technique. Radio stations, especially FM types, are becoming totally packaged,

run from computer tapes. The announcers "sound"
spontaneous. They exude a now quality; their chatter
is slick, neighborly, bright. A California station has
nearly doubled its hearers since becoming fully auto-
mated! It just *seems* more real.

Arthur Schlesinger, Jr., in his remarks to the conference on "Leisure, Technology, and Human Values,"
notes:

> I think the basic reason for the surge of disquietude which
> is seen among the young is the sense they have of being
> born into a world dominated by vast, impersonal, towering
> structures impervious to human need or human desire. This
> is a very disquieting situation and I think the struggle be-
> tween the individual and the organization is a distinctive
> conflict of the high-technology society.

This sense of manipulation and disquietude leads to
the second revolution. Before describing this we must
point to the fact that not only does technological innovation require more preparation time to learn how to
use the new techniques, but the very technology ex-
pands the amount of free time available and thus de-
creases work. Added to this is the shifting of work
away from primary and secondary production to the
tertiary levels of consumer service and social service.

Finally, these technological accomplishments have
brought unprecedented abundance and opportunity.
Those born into the society during or since World War
II have known for the most part only well-being. They
have had the good life and been given time to enjoy it.
Wealth and free time have increased to such a magnitude that a new orientation toward life is emerging. It
is for these reasons we might say we're entering the
post-modern era.

We noted a dual revolution. Coupled with the technological and scientific mutation, which has produced increased societal and environmental control, is the deliberate choice by young people to defer entry into the adult world.[4] The very affluence and technology makes this possible. Yet the reason for choosing to wait is a disenchantment with the "administrative state"[5] and the Scarsdale life-styles. The coming together of all these factors has produced a different consciousness among youth. They are in many instances marching to a different drummer. They have encountered the corporate state—the creature of the new technology and technique—the overweening bureaucracy, the objectification of their humanity and of others. They are not free to be independent, creative, to become what they have been extolled to be. They don't like it, can't accept it. They are convinced there is something more. They will and do wait before signing on the dotted line their contract to this corporate world. Such widespread disengagement has enormous impact.

The magnitude of this decision is what is creating

[4] Dramatic figures of school enrollment, from the Bureau of Labor Statistics, 1970, suggest this disengagement by youth and their deferment of entry into the labor market. In 1950, 9 percent of the twenty- to twenty-four-year-olds were enrolled in school; in 1960, 13.1 percent; and in 1969, 23 percent.

[5] Charles Reich uses this term in his provocative book *The Greening of America* (New York: Random House, 1970): "The structure of the administrative state is that of a hierarchy in which every person has a place in a table of organization, a vertical position in which he is subordinate to someone and superior to someone else. This is the structure of any bureaucracy; it represents a 'rationalization' of organization ideals. When an entire society is subjected to this principle, it creates a small ruling elite and a large group of workers who play no significant part in the making of decisions" (p. 99).

this second mutation. It is an ethical and aesthetic revolution that goes far beyond simply refusing to enter the adult world. Because it is aesthetic and ethical its meanings arise from the profoundest cultural levels. As Daniel Bell has observed, there is a sharp cleavage between culture and social structure in our society.[6] This is always an indicator of deep societal crisis auguring a discontinuity of social types.

Putting it together: free time, abundance, protracted schooling, organizational bigness and closure, environmental destruction, unprecedented violence (resulting from the corporate state being a combination of the political, economic, and military which is carrying on a war in Southeast Asia that increasingly seems to be a way to assure an open door for the insatiable hunger of the American economy), disaffection with the possibility of meaningful occupation—all these factors produce a synergetic movement of malaise, discontent, disquietude, sometimes angry and violent and sometimes escapist, as in the drug scene.[7] This is the making of the second mutation. It has moved us into "the country of the young," to use John W. Aldridge's words

[6] See Bell's fine article "The Cultural Contradictions of Capitalism," *The Public Interest*, Fall 1970, pp. 16-43.

[7] There seems to be a close correlation between free time (wanted or unwanted) and the use of drugs. Elements of youth seek to escape where they find themselves, whether it be the ghetto of Harlem or Scarsdale. Much more research must be done on the increased use of drugs. This is a growing problem. I argue that it is a lead indicator of social ambiguity marking a transitional period and a profound societal crisis, and that it is related to the discontinuity of social norms. It is striking to reexamine the two-volume set *Hearings Before the Select House Subcommittee on Education,* 1970, and note not a single dealing with this issue of "escape."

for it.[8] He decries our servitude to the young but fails to ask why we have reached this circumstance.

This disengagement is producing a new life-style. Charles Reich calls it a new consciousness. His description fits very closely our own. He calls it Consciousness III. He sees this response as "the product of two interacting forces: the promise of life that is made to young Americans by all of our affluence, technology, liberation, and ideals, and the threat to that promise posed by everything from neon ugliness to boring jobs to the Vietnam war and the shadow of nuclear holocaust. Neither the promise nor the threat is the cause by itself; but the two together have done it. . . . To them [the young], the discrepancy between what could be and what is, becomes overwhelming; perhaps it is the greatest single fact of their existence." [9]

Many argue that every generation in its youth has felt this same gap existing between principle and practice. Today's "older people," Reich states, "are inclined to think of work, injustice and war, and of the bitter frustrations of life, as the human condition. Their capacity for outrage is consequently dulled." [10] Two things are important in this statement. First, it would seem that older people's capacity for outrage has become dull because they have a vested interest in the fruits of their work. The present young have never really had to work.[11] Now they are disparaging the

---

[8] *In the Country of the Young* (New York: Harper & Row, 1970).

[9] Reich, *The Greening of America*, pp. 218, 220.

[10] *Ibid.*, p. 220.

[11] We argue that the affluence of the category Youth has freed them from work. They hold some part-time jobs, but only for the purpose of being able to support a car or a girl friend (or

very condition their elders have worked for all their lives. They "spit in their father's soup." This act sets up serious cognitive dissonance. How can older people really admit that what they have been working for all these years is really that bad?

Second, work itself is changing. This is probably the most important factor. The nature of work is shifting from an emphasis on primary and secondary roles of productivity to the tertiary roles of consumer and social service. The latter especially are increasing as programs of strategic social change are increasing. Work is no longer dominating one's life-style as it previously did. We are moving toward what Joffre Dumazedier and Max Kaplan call the "leisured society." The new orientation is on leisure. The consciousness—for which I would substitute *weltanschauung* in the sense Albert Schweitzer used it in his *Philosophy of Civilization*[12]—looks at life primarily from the angle of free time, *discretionary time*. The amount of this time has almost single-handedly created this new orientation.[13] The content of this time is culture. And a cultural consciousness emerges among the young. "Always before, young people felt themselves tied more to their families, to their schools, and to their immediate situations than to 'a generation'! But now an entire culture, including music, clothes, and drugs, began to

---

both), or to buy new clothes—in other words, to consume. As "suburbanization" increases (see the 1970 census results), the category of Youth increases, and hence, this pattern of relative disengagement. Likewise, there are fewer jobs for Youth. Unemployment among Youth is at an all-time high.

[12] Trans. C. T. Campion (New York: The Macmillan Co., 1960), p. 271.

[13] See Appendix A for Juanita Kreps's description of possible amounts of free time for the future.

distinguish youth. As it did, the message of consciousness went with it. And the more the older generation
rejected the culture, the more a fraternity of the young
grew up, so that they recognized each other as brothers
and sisters from coast to coast." [14]

This world view on the part of youth is something
new. It is a reality which is becoming extremely widespread among college-university students and college-
bound high schoolers. The studies of Richard Flacks,
Kenneth Keniston, and Richard Peterson basically see
this youth culture coming from the privileged, upper-
status levels. "Our data suggest that, at least major
Northern colleges, students involved in protest activity
are characteristically from families which are urban,
highly educated, Jewish or irreligious, professional and
affluent. . . . Mothers are uniquely well-educated and
involved in careers, and that high status and education
has characterized their families over at least two generations." [15]

Likewise Richard Peterson finds that " 'student
leftists' are upper middle class in their social origins.
Their parents are politically liberal or radical, and
many have been involved in radical politics. Both the
students and their parents consider themselves to be
either nonreligious, or liberal and nonformalistic in
their religious orientation. Parents of activists are
permissive and democratic in their child-rearing practices, and their children are both highly intelligent
and intellectually rather than career oriented." [16]

[14] Reich, *The Greening of America*, p. 224.
[15] Richard Flacks, "The Roots of Student Protest," *The
Journal of Social Issues*, XXIII, 3, 52-75.
[16] Richard Peterson, "The Student Left in American Higher
Education," *Daedalus*, XCVII (Winter 1968), 304.

Kenneth Keniston notes that the young group he studied by and large came from families where prosperity and luxuries were taken for granted. Their families were "generally free from acute anxiety over status;"[17] their parents were thoughtful and well educated.

Though this may have been the case—and perhaps generally still holds—the *influence* of these upper-status young protesters, dissenters, critics has been much more pervasive. In this sense I agree with Charles Reich that there seems to be a revolution underway. A new world view (consciousness, he calls it) is coming into sight.

There is a revolution coming. It will not be like revolutions of the past. It will originate with the individual and with culture, and it will change the political structure only as its final act. It will not require violence to succeed, and it cannot be successfully resisted by violence. It is now spreading with amazing rapidity, and already our laws, institutions and social structure are changing in consequence. It promises higher reason, a more human community, and a new and liberated individual. . . .

This is the revolution of the new generation. . . . It is both necessary and inevitable, and in time it will include not only youth, but all people in America.[18]

A key notion here is that the source of this new order in the making is from the cultural levels. Profound societal change always means a basic transformation of fundamental norms. The sacred is overturned, in fact dies, and a transfiguration occurs. What rises from the ashes is a new normative structure providing a new ordering of values, that is of the quality

[17] *The Young Radicals*, p. 230.
[18] Reich, *The Greening of America*, p. 4.

of *being*. We so often view revolution only from its political and economic manifestations. These are always the last to change. The real sources of change reside with culture.[19]

A second notion in Reich's statement I seriously question. He leaves the general impression with the reader that this process is inevitable and will become increasingly widespread. Though he uses the conditional tense and appears to limit the "necessity" of the "greening of America" by the new generation, he seems to believe that it will happen, that we need only step back and wait. The movement is underway and unstoppable.

I take issue with this implication. The quality of the individual and collective human psyche does not allow for such a certain determinism, i.e., inevitability. There are too many variables, and there is a will-o'-the-wisp quality to the human spirit which defies such certitude. Moreover, to assert inevitability is to reduce human freedom and choice to nothingness. Thus, the forces of fixity and persistence may be powerful enough to stop such a development, or the mutation which will come will be very different from what Reich envisions.

Still where does that leave us as to prediction? Can we determine any trends for the next decade? I think we can. There seems to be a very broad cultural movement causing the changes now appearing.

From another source we find agreement. Daniel Bell,

[19] For a discussion of the cultural, normative sources of societal change, see E. A. Tiryakian, "A Model of Societal Change and Its Lead Indicators," in *The Study of Total Societies*, ed. Samuel Z. Klausner (New York: Doubleday & Co., 1969), pp. 69-97; and Georges Balandier, *The Sociology of Black Africa* (New York: Praeger, 1970) and "La sociologie des mutations" in *La Sociologie des mutations* (Paris: Anthropos, 1970).

writing in *The Public Interest*, asserts that "culture
has been given a blank check, and its primacy in gen-
erating social change has been firmly acknowledged." [20]
He goes on to say: "The social structure today is ruled
by an economic principle of rationality, defined in terms
of efficiency in the allocation of resources; the culture
in contrast, is prodigal, promiscuous, dominated by an
anti-rational, anti-intellectual temper. The character
structure inherited from the nineteenth century—with
its emphasis on self-discipline, delayed gratification,
restraint—is still relevant to the demands of the social
structure; but it clashes sharply with the culture, where
such bourgeios values have been completely rejected." [21] Finally, he suggests that "contrary to
Marx's idea of culture 'reflecting' an economy, inte-
grally tied to it through the exchange process, two distinct and extraordinary changes are taking place. Art
has become increasingly autonomous, making the artist
a powerful taste-maker in his own right; the 'social
location' of the individual (his social class or other
position) no longer determines his life-style and his
values." [22]

This is a crucial observation. It agrees with what
we were saying before. The segment of youth appearing in the Flacks-Keniston-Peterson studies came from
a certain social stratum. However, the decade of the
60's changed that. The influence of these youth cultural leaders has so broadened that those participating
in these cultural movements come from all strata.
Hence, the new life-style may be appropriated by anyone from any kind of background anywhere. "The ques-

[20] Bell, "The Cultural Contradictions of Capitalism," p. 18.
[21] *Ibid.*, pp. 18-19.
[22] *Ibid.*, p. 20.

tion of who will use drugs, engage in orgies and wife-swapping, become an open homosexual, use obscenity as a political style, enjoy 'happenings' and underground movies, is not easily related to 'standard variables' of sociological discourse." [23]

A parallel crucial aspect of Bell's analysis is that *culture* is a way of life and leisure dominates the new culture. Leisure is becoming a way of life. To choose life-styles implies having discretionary *income, time,* and hence *social behavior*. People—youth especially—increasingly want to be identified by their life-style and cultural taste, rather than by occupation. Technological and scientific advances have made possible these three discretionary features. Their intersection is creating a new world perspective, a new consciousness, which is a hallmark of the "leisured society." Discretionary income can be spent for those items not considered basic needs; discretionary time means time free from work; discretionary social behavior opens a myriad of life-styles to anyone regardless of family, education, age, wealth, ethnic background, and location. This is a new type of society.

It is a perspective based on openness, choice, fluidity, flexibility, change, and spontaneity. It's main attribute is innovation, inventiveness, the new. "Our culture has an unprecedented mission: it is an official, ceaseless searching for a new sensibility." [24] The few may be exploring the new but "its behavior-styles diffuse rapidly, transforming the thinking and actions of larger masses of people." [25]

The White House Conference on Youth, 1970, had to

[23] *Ibid.*
[24] Bell, "Unstable America," p. 17.
[25] *Ibid.*

take into consideration this new orientation and world view. Youth, as one of the first groups created by the post-industrial technology and science, is the vanguard. High school and college-university students are the important groups of this newly conscious entity. Students in higher education number over 8,000,000. High school enrollment is about 18,000,000. The average length of time in school for everyone is about 12.1 years.

The post-industrial life-styles of youth are diffusing rapidly with high schools and junior high schools. In many ways it is a conscious effort. Some self-anointed of the new age take their way of life directly to the secondary schools with a missionary zeal.

The page appearing as Appendix B was distributed throughout high schools and junior highs in the Tampa Bay area. A real hassle then developed between the university and the community. Drugs, mostly pot, are widespread at events like these. The emphasis is on a celebration of **life,** development of sensitivity, openness, expressed through touch, sexual experience, self-expansion. Eastern religions are popular, yoga, and cultic religious experiences.[26] Surveys by my students in a *social change* class show the rise of such activities related to the renaissance of an interest in the inner or affective side of life; "I'll be all right, when my head gets right" is a common way to express this.

My undergraduate counseling load is average in size and in type of student. Yet four of six students who recently sought advice on their sociology program indicated they would either stay in school, travel around the country, or leave for Canada to stay out of the organized, corporation society. They could not accept

[26] See "Human Potential: The Revolution in Feeling," *Time,* November 9, 1970, p. 54.

what was waiting for them in the adult world, translated "the corporate world," the "administrative state," "Scarsdale," or what have you. How large this group is I don't know. We have no studies. Their influence is considerable, however. That is the crucial thing I'm suggesting here.

I found in reading the materials from the White House Conference on Youth, the Task Force on Economy and Employment, that the emphasis did not take these "non-institutional" factors sufficiently into account. The consultants seemed oblivious to these influential cultural levels. This brings us to the question: why is it important to consider *youth* as *the emerging* group of the post-industrial age?

1. Youth are deferring longer to enter the work force.[27] Quantitatively, their skills and leadership are important economic factors, which increasingly are unavailable.

It is comforting to argue that they will make up their minds as they gain responsibilities of family and grow older, hence wiser. My point has been that *youth* are different from *adolescents*. The argument of a "natural" conformity eventually occurring is inappropriate. Youth see things differently than adolescents. They have a different ordering of norms (values). They are seeing the family differently, and hence, family responsibility. They will enter adulthood eventually but on their own terms.

In summary, the following factors seem to influence youth to defer entry into the work force: (1) Their

---

[27] The statistics on education illustrate that growing numbers of youth are choosing to stay in school rather than go into the labor market. I define school as a free-time (leisure) activity.

education has become increasingly protracted while at the same time it is becoming a leisure activity. (2) Youth's affluence has enlarged steadily which gives them the resources for mobility and the choice of life-styles. (3) They have experienced an incredible array of events which, since World War II, have produced the fastest rates of social change in man's history. The meaning of work, among other things, has changed. (4) Violence is everywhere and is incongruent with the principles youth have been taught, to wit: the Vietnam war described every evening on television, the violence to and waste of the environment, the slaughter on the highways by vehicles too powerful for a society to handle, the assassination of beloved leaders, the murder of minority group movement leaders, and the denying of basic freedoms to poor people such as Mexican-Americans, blacks, Indians, and Appalachian coal miners. Youth have been told of the American dream. They have saluted the flag every morning in school since they can remember. They have heard Fourth of July speeches and their parents talk about fighting wars for democracy, for humanity. They sense there is too wide a gap between what they were told and what they see and experience. How can they accept such hypocrisy? "By God, I won't," some cry. They call the culprit the "system," the "establishment," the "man," the "economy," the "society." They are over against whatever it is; they can't join it. Paradoxically, ironically, they have the affluence, they can work part-time at odd jobs sufficiently to defer entry for a considerable length of time!

2. Their views on society and life-style are not isolated, contained, or constricted. The very technology which has freed them, given them the options in time,

life-styles and income, provides the rapid diffusion of their patterns of culture. This is a part of what Marshall McLuhan is saying in *Understanding Media*.[28] From the standpoint of political action, a CBS survey noted that the majority of students agree with the SDS objections to, and criticisms of, the American system.[29]

If one examined that same CBS survey for attitudes on basic features of the American way of life as perceived by students, it is obvious how widespread are these feelings of malaise, disenchantment, even disengagement. CBS used the University of Colorado at Boulder to film the report on the survey since it contained such a representative spectrum of the attitudes which appeared in the survey. Colorado is no radical, Northeast, Ivy League school! CBS could have selected the University of South Florida just as well.

3. The White House Conference sought to make projections and policy suggestions for the 1970's. My contention in this paper is that the orientation of youth described above will dominate the decade. The policy suggestions hence must reflect this reality.

The emphasis on employment for the poor and on other economic solutions for people in poverty, especially the plight of minority groups, is important. However, to see the attitudes of all youth as being basically oriented toward such quantitative, economic goals is too narrow. There is a large segment—as yet statistically difficult to pinpoint—which is looking at the world through qualitative lenses, or, in Daniel Bell's analysis, through the optics of culture. From my own

[28] (New York: New American Library, 1964).
[29] *Generations Apart:* a study of the generation gap conducted for CBS News by Daniel Yankelovich, Inc., 1969. See the section on "Institutions and Restraints," beginning on p. 17.

perspective, through the lenses of leisure, defined in a broad, general sense. Large segments of youth are disengaged from the economy. They move in and out of jobs just to get "the bread." They have no long-term commitments. They are reluctant to make any. They are looking for something *meaningful.* They want to serve, to be useful, to be human. They see the corporate state, "the system," basically opposed to such goals. Their very criticisms require a structural change of the society, a reordering of priorities. Such a profound structural transformation seems in the offing, supporting our contention that we are on the edge of a new societal type.

4. Another reason why youth must be seen in the 70's as the vanguard of the post-industrial society is because their views, their orientation on the worlds of time, life-styles, and the economy force us to look at these issues differently.

Life (not work), in order to be meaningful, must contain elements of service and/or creativity. The situation of youth which gives them discretionary time, income, and life-style reintegrates the spheres of work, education, religion, etc. This is the nature of the leisured society. Human fulfillment for both the actor and those for whom and with whom he acts must be the conscious goal.

Human growth is to be substituted for crude economic growth. Hence, qualitative meaning will come, among other ways, from involvement in projects and programs of social amelioration. The plight of the poor takes on a special priority. Youth, seeking something beyond the rat race of the marketplace, find meaning to their work and function when engaged in such a labor of amelioration. That meaning comes through the renewal of hope on the part of the underclasses of this society. Other social problems which threaten our survival and well-being (human values) need similar attention, e.g., peace, racial conflict, environmental destruction, population, etc. If youth could not actively engage in such programs because they lack sufficient skills, they could be preparing themselves for such *activity*. The present orientation of the economy is not committed to developing strategic programs on the massive scale the problems require for solution. Nor does the economy provide these kinds of options. If you are interested in making a profit, if this is the source of your kicks, then the economy has some validity for you. If, on the other hand, you seek other goals, seek meaning in human welfare and fulfillment, the present corporate structure provides few outlets. Any structure that feeds on the unremitting destruction of resources, land, lakes, sea, through constant speculation and "development" as it is manifested in Florida, cannot fundamentally, normatively be concerned with human fulfillment and the human community. Any system that can systematically, massively destroy people and land and resources as ours has done in Vietnam (now all of Indochina) is disinterested in the well-being of the human person.

Ironically, youth can defer entry into the system because the very affluence of the system enables them to tap it sufficiently to stay comfortable. They are ambivalent toward it in this sense. They fear and hate it, but use it to disengage. The magnitude of their disengagement is changing the system's structure and orientation. This is our point. Where it goes, what that ultimate orientation will be is not certain. The trends seem to be in the direction of the new life-styles, "patterns of culture," identified with the young. It is on the basis of these trends in youth cultural patterns that the following policy recommendations are made.

Policy Recommendations

The orientation of youth as an emerging group of the post-industrial society seeks new values for living. Youth find a spiritual emptiness to affluence. As we noted, they have no real need to strive for status or subsistence, or security. They have these. What replaces these quantitative factors? Qualitative concerns. What are these?

1. Youth desire opportunities for personal development. They desire their own human development to be lifelong, without fixity, without an ending. This obviously has immediate and long-range implications for careers themselves, i.e., employment.

2. Youth desire opportunities to help other persons realize their human potential. Hence, youth seek careers to aid poor people wherever they are; to engage in service careers, such as teaching, creative arts, public health, conservation of land through the development of parks which relate man and his natural environment, careers which attack the problems of the city, the third world, race and ethnic relations, etc.

3. Obviously, it is difficult to separate careers and career preparation. Preparation has to be *for* something. If the career opportunities are dead ends, then the preparation will be meaningless. If the career (vocation—calling, in Latin) is *life, being, fulfillment,* then preparation (education) must be directed toward these goals.

The economy's task is to allocate resources for this purpose. The openings and options must be there. But first, the goals of the economy must be toward *human fulfillment,* not control, repression, destruction, opportunism, or "let the buyer beware." In other words the whole system must strive toward the goal of human growth.

I'm talking here about a basic, fundamental philosophy or orientation toward the economy. The economy as an institutional arrangement should serve the community and the human person, not destroy them or beguile them. Christmas, the annual Western industrial world's potlatch, has become a time during which the public is cajoled, bombarded, enticed, and propagandized into buying (consuming) great quantities of things to express their supposed love for one another. The credit card becomes "Santa's helper." Debts accumulate to fantastic levels. Parents feel guilty if they can't match their neighbors' capacities to heap great mounds of "presents" on their children. The potlatch starts earlier each year. We barely finish Halloween before Christmas decorations are up in the self-service department stores. It's all symptomatic of the real goals of the economy. No wonder youth increasingly call for simplicity, seek to be self-sustaining in communal groups (now over six hundred such communities in the United States), and seem so thoroughly disgusted with the life-style of Scarsdale. I suspect the rapid rate of increase in shoplifting, especially among youth, is related to an economy which is based on creating false needs, which places so little value on the human person, and whose philosophy is to make a profit and "let the buyer beware." Nader's Raiders' studies and exposés show clearly how true this is. Manufacturers' guarantees are meaningless; "enriched bread" is valueless nutritionally; cereals contain little that has nutritive worth; the aged are being increasingly segregated and taken advantage of, especially in so-called convalescent or nursing homes. It's an economy without principle. Its sole *modus vivendi:* make money, grow. It tries to tell us everyone will *profit* (note the choice of words) if the economy grows. Our incomes may expand, but so does the cost of living, and so does the wanton destruction of resources, landscape, human lives, and communities.

My point is: "This killing must stop." How can we go on preparing, training, educating persons for the corporate state and economy? You don't know really how bad it is until you ponder Vietnam! A large, influential segment of youth are saying, "We can't enter such an economy." We want our lives to count. We want to build up humanity, not destroy it. How can we do this? How can the economic system be reworked so it is moving in the direction of human service? This is a revolutionary question and a revolutionary task. It is no ordinary request. It goes beyond reform. It is asking for a totally different economic solution, substituting quality of life for crude growth. This coincides with the emergence of the leisured society, which is a society whose main focus is on the quality of life, on the re-creative, on structures which free man to flourish as a human being. The search is for these structures.

By revolutionary task we do not mean it will take a bloody uprising to accomplish the goal. In many respects—and this is the irony—the Vietnam war is doing it. The existence of 30 million terribly poor people in this country is doing it. The repressive tactics of the growers against the migrant laborers and the murder of black leaders, especially Black Panthers, are doing it. The involvement of the military in surveillance of private citizens, an activity which has no relationship to national defense, is doing it. The slaughter on the highways because we've been cajoled into thinking that big engines in cars make males more sexually attractive, or something like that, is doing it. The killing of Lake Erie by industrial pollution is doing it. The ethical-aesthetic response to this violence gives hope and cause for optimism. It is a revolutionary response. The opting for different life-styles, uses of incomes, and employment of time are beginning to create the big change. Because it comes from the cultural levels means normative change is underway. That is why we would call it revolutionary in scope.

Specifically, recommendations for career preparation have to do with a whole new concept of education, based not on consumption and production and profit, but on service, use, and re-creation. Ivan Illich would say that schools have to be replaced by education. Schools are geared to turn out consumers and producers. Education seeks human actualization. It's an important distinction.

This education means a place and a structure wherein one finds his own level, his own interests, and measures himself. (What does this *mean* to me? How do I apply this to myself and my situation?) It is an intimate search for truth in the give-and-take of dialogue, allowing one's curiosity to flower, learning from one another. It means learning in the human community, dispersed, decentralized.[30]

The above kind of preparation is essential for persons to function in terms of service, adaptability, seek-

[30] See George Leonard, "Education and Ecstasy," and Ivan Illich, *De-Schooling Society.*

ing to develop their own personal potential, and to improve the community. This returns us to the theme of leisure. In the Greek notion of leisure—which for them is a state of mind—one seeks in and through his being to express himself to the fullest of his potential. He does this in creative acts, above all, in contemplation and other mind-expanding, spirit-expanding activities. There is no sense of occupation to leisure. It is the end goal of existence. It is true being itself.

There is another side to this notion of leisure for the Greeks. That is, not only does being imply self-growth, self-actualization, but it also means the building up of the community, the human polis, with the aim of providing the structures supportive of leisure. Hence, the ethical responsibility of the individual is to himself and to his community. Neither can exist without the other.

How should a person prepare for such leisure? We are talking basically about a new orientation of service, of upbuilding (reconstructing) the community and of self-growth. The two are intertwined. Careers for this purpose are different from those of the industrial society. They no longer are production-consumption directed. They are service-oriented. The preparation must be different as well. Hence, these recommendations follow from these perspectives of the leisured society.

Types of careers. For the category of youth there seem to be two broad types of service careers, short-term and long-term.

Short-term Careers. I recommend that:

1. The task force go on record as supporting the newly established International Peace Corps and urging that the United States Peace Corps be dissolved. This

fits a necessary philosophy of foreign relations based on multilateral relationships rather than bilateral ones.

2. Private and/or university agencies be encouraged to work out Peace Corps type programs, to support these efforts with funds, and to combine education and work in the field.

3. Great support be given to Vista.

4. Private and public agencies be strongly encouraged (through funds) to develop Vista type programs. A part of the "American genius" is its pluralism of projects and programs. Individual initiative to seek service through such organizations should be encouraged in a direct way.

5. Two types of blanket programs be tried, or certainly at least explored. First, that the creation of the National Service Corps which is before Congress be supported. Here males and females (eventually) when draftable have the option of choosing the military or an existing service program, or working out a project of their own for approval. The options would be open only in time of relative peace.

Second, the work/study plan of several schools and colleges be explored. An alternation is set up whereby a person goes to school for a quarter or semester and then works in an area of interest for a like period of time. Some universities have combined Peace Corps and graduate-study experiences.

6. Encouragement be given to individual growth projects such as a six months' motorcycle tour of either this continent or the third world, combining work (not panhandling) with travel.

7. Many kinds of projects be supported: service opportunities on staffs of halfway houses, in centers like Wilmette House, Cambridge, Massachusetts, or Penn

Community Services, Frogmore, South Carolina; neighborhood service centers, model cities projects, poverty programs, human relations commissions, Nader's Raiders, psychiatric hospitals, park services, recreation services. Massive inputs of money and project development at the federal level are required for these to be effective.

Long-term Careers. I would suggest that:

1. Anything over two years is long term. This is another hallmark of the leisured society. Presently the average American holds twelve different jobs during his lifetime. This moving in and out of programs will increase. Currently the Peace Corps requires an administrator to leave the agency after five years. Alvin Toffler notes in his *Future Shock* that "adhocracy" will be the pattern of administration of the future. Task forces like this one will increasingly attack specific problems and solve them. The task force will dissolve after the task is complete. Here learning must be considered.

2. Support of the notion that learning move increasingly into the environment. In this way, the university and high school disperses, decentralizes. Hence, study/work projects could alternate on quarter cycles, yearly cycles, two-year cycles, seven-year cycles, etc. The criteria must be flexibility and options. The leisured society is a society of discretionary time, income, and life-styles. The industrial society is one of uniformity, standardization, homogenization.

3. Careers of long term would be doing many of the same things as short-term ones. I would recommend giving priority to such a problem as conflict resolution both within the nation and on an international level.

Thus, community human relations programs need to be drastically strengthened, upgraded, expanded.

4. The UN needs the unqualified support of the United States, especially in their work through UNESCO, Economic and Social Development, World Bank, UNICEF, World Health, Food and Agriculture, etc. These careers in the UN type organizations should be expanded, made attractive, given utmost priority.

5. An implication through all this is some sort of guaranteed annual income. The leisured society is one in which persons have options, in which the Greek ideal of leisure may be possible, i.e., individuals and groups seek self-fulfillment and self-growth and, at the same time, seek to reconstruct their communities wherein such human fulfillment may take place. I suggest the task force go on record as supporting this notion from this perspective. Such a goal has already been expressed. This is an added dimension and justification for the proposal.

6. Institutional imagination is required to break through the old patterns of work, weekend, vacation, and work, to retirement—and retirement so often equals being "put out to pasture." The present abundance reflected in the Gross National Product, making possible a guaranteed annual income, leads to the notion of *Committed Spending*.[31] This means the decision by society to support each person in terms of a flexible use of work and nonwork time. Such a committed spending needs to take into account the differences in occupation *plus* individual preferences as well.

---

[31] See Robert Theobald, editor, *Committed Spending: A Route to Economic Security* (New York: Doubleday & Co., 1968). This book contains some excellent articles reflecting institutional imagination.

Again emphasizing options, each person should have the opportunity to work out his own pattern of work and leisure. Hence, a large number of options should be available. What are possible options?

Daily

½ days, 5 days a week
½ nights, 5 nights a week
2 days, 2 nights each week

3 days a week at 12 hours each day
4 days a week at 10 hours each day

Or any other combination. It depends on the job and the employee.

Weekly

3 weeks each month
6 weeks on and 3 weeks off at 40 hours each week

Or any other combination. Again it depends on the nature of the work and the personal needs of the employee. Actualization is the goal.

Monthly

6 months on and 6 months off
3 months on and 3 months off
2 months on and 2 months off
8 months on and 8 months off

Any combination would be feasible, again considering the job.

Yearly

2 years on and ½ year off
5 years on and one year off

Again, any combination is conceivable.

These are not impossible suggestions. A good computer programmer could work out such combinations for an institution like the university with 1,000 employees.

RESEARCH INTO LEISURE

ALEXANDER SZALAI

THE 12-NATION COMPARATIVE STUDY

Many interesting patterns of social life are associated with the temporal distribution of human activities, with regularities in their timing, duration, frequency, and sequential order. Certain techniques of data collection based on direct observation, interviewing, and the examination of records permit the establishment of fairly adequate itemized and measured accounts of how people spend their time within the bounds of a working day, a weekend, a seven-day week, or any other relevant period. Investigations of this particular aspect of social life based on the quantitative analysis of such accounts are commonly referred to as "time-budget studies."

This conventional name has some metaphoric justification, since very many studies of this kind are concerned primarily with the proportions in which the twenty-four hours of the day are allocated to various activities by people belonging to certain groups or strata of the population—how many hours and minutes such people spend daily on chores and pastimes such as doing work, putting things straight in the household, shopping around, taking meals, visiting friends, reading books, listening to the radio, having a good night's rest, and so forth. This type of analysis is indeed somewhat similar to the procedure by which the allocation of funds for different purposes is reported in financial budgets. As far as personal or family budgets are concerned, the similarity will even extend to many specific types of expenditure, because a great number of everyday activities involve not only the expenditure of time, but of money as well. At this point, however, the resemblance comes to an end. Time can only be spent, not "earned." Therefore, time budgets have no income side. The fund of time which is being "allocated" to various activities (24 hours in daily time budgets, 7 times 24 hours in weekly time budgets, etc.) serves simply as a frame of reference for setting out the temporal proportions of people's engagement on the whole gamut of their daily activities. Thus, it is not time itself, either as a physical or as a subjectively perceived entity, but rather the use people make of their time which is the real subject of time-budget studies.

The use of time consists, of course, in doing something, in conducting some kind of activity. Even loafing and resting are to be understood as activities in this context. Studies on the use of time are thus, in effect, concerned with the daylong activities of people and, more specifically, with patterns of social behavior that can be revealed by registering certain temporal attributes of individual and collective human activity as it unfolds in the course of daily life.

Time-budget studies of the traditional type take account mostly of a single temporal attribute of human activities, namely, their duration. So many hours and minutes spent daily in sleeping, working, eating, traveling, housekeeping, reading, TV viewing, etc., by a person belonging to a given group or stratum of the observed population—this is the typical format in which data are presented in traditional time-budget studies. For obvious reasons, the data pertaining to working days and nonworking days may be put forward separately; also a more or less refined breakdown according to types of activities may be applied. Nevertheless, the essential structure of the time budget remains always the same: a "block" of time (an aggregate amount of hours and minutes) that represents the per capita average of the total daily time expenditure by the group on a specific kind of activity is being tacked on to every item on the daily agenda of people, and these blocks add up to the 24 hours or 1440 minutes of the day.

For some limited purposes, even such aggregate average daily duration data can prove quite useful if carefully collected and analyzed. For instance, planners of communal services and facilities may use them for determining certain gross parameters of the popular needs their establishments have to meet. Also, by comparing the relative proportions of time allocated to various activities by people in different walks of life, some insights can be gained into differential living conditions, social interests, or cultural preferences prevailing in certain parts of society.

Nevertheless, time-budget data of this kind, taken alone, lack the necessary depth for any more penetrating analysis of social behavior. First, the aggregate time devoted daily to a certain type of activity gives no information about the frequency with which that activity is undertaken during the day or about the typical duration of its single occurrence. It is, in many respects, far more interesting to find out how often a man switches on his radio during the day and how long he is typically listening each time than just to know how many hours and minutes he spends in listening on the whole. Second, apart from a few functions fulfilling direct physiological needs like sleeping, eating, or performing certain acts of personal hygiene, there are surely not many types of activity which we would expect all people to carry out at least once each and every day. If we learn that within a group of people a per capita average of eight minutes is spent daily on reading newspapers, we still do not know anything about the proportion of people who effectively read a newspaper on any given day, nor do we know how much time is devoted to newspaper reading by those people who do in fact indulge in this particular kind of activity. In short, aggregate average daily duration data as described above disclose nothing about the typical duration of the individual act—neither the typical frequency with which it is being performed during the day, nor the proportion of its "doers" in the observed population on an average day.

Another deficiency of a very considerable part of time-budget studies arises from the fact that individuals often perform several activities at the same time.

In particular, acts of active and passive communication, like talking, listening, watching, reading, etc., tend to be carried out while doing something else. People carry on much of their daily conversation as they take meals or do some kind of work that does not engage fully their attention; they read newspapers while commuting to their working place and listen to the radio while preparing their breakfast. Therefore, in order to produce time budgets in which the total duration of daily activities adds up to 24 hours or 1440 minutes, one has to choose the one activity to be counted whenever a person has carried out several activities simultaneously. Such an elimination of secondary or parallel activities from the circle of observation naturally distorts in a rather arbitrary fashion the picture of what people do the day long and leads to a biased account of the amounts of time they devote to various tasks of life.

Admittedly, there are practical limits to the accuracy with which the many different, parallel, and criss-crossing threads of activity can be followed up in the complex fabric of everyman's daily life. Moreover, for whatever level of accuracy one may reach, still more minute observations could possibly prove that some activities which seemed to be carried out simultaneously were in effect alternating with one another, or that some activities which seemed to be performed consecutively were factually overlapping to some extent. Nevertheless, any time-budget study which does not grapple in some way with the problem of recording secondary or parallel activities is essentially unable to give a balanced account of the great variety of activities that fill up everyday life. This is especially true of studies on leisure, a subject to which the major part of traditional time-budget studies has been devoted.

At least two of the most widespread leisure-time activities in our days, listening to the radio and viewing television, are enjoyed to a considerable extent simultaneously with other pastimes. People quite often tend to relax by doing nothing specific, i.e., letting themselves go in various ways at the same time: family chit-chat while watching television and sunbathing on the beach while listening to the transistor radio and watching girls are just a few examples. People also enjoy stealing some leisure while working very hard at the same time: they smoke, they sip coffee, they doodle, they hum tunes, and even listen to music while bending over their desk or working bench. At this point, however, we are already approaching depths of human behavior no survey research can hope to reach. Surely, time-budget studies cannot attempt to compete with James Joyce's *Ulysses,* though their subject is much the same: a typical day in the life of a very common human being. But this should be no excuse for reducing the manifold and often simultaneous activities of individual life to a single chain of events, the duration of which adds up to exactly twenty-four hours per day.

Even for the purpose of time-budget studies concerned only with the average duration of various daily activities, basic data have to be collected, mostly by registering events in the daily life of individual respondents in their temporal order. Without some kind of "historical" report on the day's activities, be it a written diary or a verbal account, it is hard to check whether any important activity has been forgotten and whether the chain of reported events really spans the whole day. Moreover, the duration of particular activities can often be assessed only by fitting them in between such consecutive mileposts in time as the ringing of the alarm clock, the whistle of the morning train, the beginning of office hours, lunchtime, closing time, the seven o'clock news bulletin, bedtime, and the like.

It is therefore somewhat surprising that so very few time-budget studies of the past have added an analysis of the timing of daily activities to the analysis of their duration, although the necessary data must have been near at hand in most cases. As a matter of fact, some highly characteristic features of social life in the industrial age are connected with timing: office hours, rush hours, shifts, timetables, and radio and television programs are some obvious examples. There can also be little doubt that the ticking of the clockwork of great society has a considerable influence on patterns of social behavior by determining to a great extent what could and should be done, or left undone, at any given hour of the day and of the night.

True, all sorts of governmental and communal authorities, public utilities, industrial and commercial enterprises have occasionally carried out investigations concerning the timing of some specific activities of direct interest to them. The daily rhythm of traveling, transport and shopping activities, or of phone calls, for that matter, has been observed and examined many times and under many conditions. However, the timing aspects of the whole course of daily life have been rarely reviewed in context. The small number of major studies of which we know in this field have been initiated by some huge radio and television companies (such as CBS in the United States, BBC in Great Britain, NHK in Japan) which had a professional interest in finding out what parts of the population are engaged in what kind of activities at different hours of the day or the night. Answers to this question have their obvious importance for determining the size and the composition of the potential audience that can be reached at any given hour, for putting the right programs (and the right commercials) in the right "time slot," etc. Regrettably, even these very valuable and methodical surveys were mainly concentrated on factors relevant to the analysis of listening and viewing habits and did not present their findings in a way which would be most helpful to the behavioral scientist interested in more general and variegated problems. Thus, much remains to be learned about timetables of everyday life in industrial society.

We have now spoken at some length about problems of duration, frequency, and timing, but we have mentioned only passingly one further important temporal attribute of human activities, namely their sequential order. While timing refers to the point in time at which some activity is undertaken, sequential order refers to the position such an activity has in relation to those activities preceding and following it. We know from everyday experience that there is some "logic" in the way people carry out their activities, one after the other, during the day. We also know that the choice of what to do next is often strongly influenced by what one has done before and what one wants or has to do afterwards. We know, finally, that some rather interesting and culturally relevant habits may express themselves in such typical sequences of activities as doing exercises after jumping out of bed, taking a nap after lunch, playing with the children before dinner, or reading before going to sleep. Little if anything do we know, however, about internal motivations and external constraints or contingencies governing sequential behavior of this kind.

Although diaries reporting in sequential order about daily activities of individuals have been collected and processed by the hundreds and thousands for all sorts of time-budget studies carried out over the last few decades, we do not know of a single published study which attempted to look deeper into the sequential structure of people's daily agenda. Of course, this kind of investigation lies beyond the horizon of conventional time-budget studies. Furthermore, the search for typical sequential patterns, their heuristic evaluation, and the assessment of their predictive value require such highly sophisticated data-processing techniques that much pioneering work will be needed before this promising field of study can be duly exploited.

All the attributes of daily activities mentioned so far—duration, frequency, timing, sequential order—are of a temporal character, i.e., involve measurements along the time axis. Quite obviously, active human behavior also has other dimensions apart from the temporal one. Researchers involved in time-budget studies have, for instance, long felt the need to broaden the scope of their investigations by including the spatial or locational aspect of everyday activities in their observation. They were met halfway by ecologists, students of urban problems, regional planners, etc., who were primarily interested in characteristics of the spatial distribution of daily life but who realized in the course of their studies that more satisfactory insights could be gained by also taking into account some temporal characteristics of the investigated phenomena. There already exists a not inconsiderable amount of research literature on combined temporal-spatial patterns of people's daily activities.

The spatial or locational aspect is of special relevance for studies in social behavior, because people tend to behave differently depending on where they are. Even outwardly similar activities often mean something very different when carried out in different places. To meet friends at home is not the same as to meet them in a club or in a restaurant; to have a private conversation in the street is something else than having it in one's own living room. Even the mere proportion of time spent during the day in various locations—indoors and outdoors, at home and around home, in streets and public places, at the workplace, in various locales and establishments—is highly characteristic of people's way of life.

Of perhaps still greater interest is to learn about the company people have during the day: how much of their time they spend with their family, with neighbors, with friends and colleagues, in the midst of anonymous crowds, or simply alone. It seems to be almost trivial to add that the character of persons interacting or even merely present determines to a very great extent what people can do and choose to do at any given time. Nevertheless, very few studies on the use of time carried out in the past have taken account of this social dimension of everyday activities, i.e., the question, With whom do people spend their time?

Who does what (and what else simultaneously) during the day, for how long, how often, at what time, in what order, where, and with whom—these are roughly the facets of active social behavior in everyday life that we have touched upon. All these facets are more or less accessible to the data-collection techniques employed in time-budget research, provided that this type of research is understood in that wider sense we have tried to expose above. More modern studies of this kind differ from traditional ones mainly in going far beyond the mere "budgeting" of durations. However, as far as we know, none of the studies preceding the Multinational Comparative Time-Budget Research Project reported in two volumes has attempted to cover simultaneously all or most of the temporal, spatial, and social attributes of everyday activities to which we have referred.

On the other hand, some special studies have included further interesting dimensions while leaving out some of those mentioned. Students of economic consumption have become interested in how people spend their money during the day; they have studied, for instance, the specific costs per unit of time pertinent to various recreational activities and the impact of such costs on the choices people make between alternative possibilities for the use of their leisure time. The correlations between the financial budgets of households and the time budgets, of whole families (taken as a group) have been investigated too; we know also of some studies about the coordination of the daily schedule of activities between family members (husband and wife, parents and children). In studies on radio-listening and TV-viewing habits the genre and content of information absorbed are often registered together with the time spent by individuals in listening to, or viewing, various programs. And so on, and so forth.

A huge amount and a very great variety of time-budget data have been accumulated over the years by many researchers in many countries. Still, little or no effort has been made to develop some more generally applicable methodological standards of time-budget research so as to enhance the possibilities for a comparative evaluation of such data. This topic came up for discussion at the international Conference on the Use of Quantitative Political, Social and Cultural Data in Cross-National Comparison, held in 1963 at Yale University. Although some new approaches to the differential evaluation of time budgets for comparative purposes could be proposed, it became quite clear that, in view of the widely divergent methods of data collection, registration, coding, classification, etc., used by researchers belonging to different countries or different schools of thought, practically no comparative method, new or old, could be made workable until at least some agreement and cooperation were achieved among a number of researchers in various parts of the world who have an active interest in this field. The idea was promoted that by carrying out a series of well-standardized time-budget surveys in different countries such agreement and cooperation could well be promoted; furthermore, a stock of standard, multinational time-budget data would be created which might provide the basis for the further development and refinement of time-budget research methods for comparative purposes and otherwise.[1]

This was one of the major considerations which led to the initiation of the Multinational Comparative Time-Budget Research Project.

The time-budget approach was developed first in social surveys reporting on the living conditions of the working class. The long working hours characteristic of early industrial development, and the struggle that

[1] See *Comparing Nations: The Use of Quantitative Data in Cross-National Research,* ed. Richard L. Merritt and Stein Rokkan (New Haven and London: Yale University Press, 1966), pp. 193-94, 239-58, 554, 569.

organized labor led from its very beginning for the shortening of the working day make it fully understandable that the proportions of work and leisure in the daily life of laborers became a matter of considerable public concern in all countries where industrialization was in progress. The famous "3 x 8" password of labor movements around the turn of the century, claiming eight hours of sleep, eight hours of work, and eight hours of recreation as the rightful daily schedule of all toiling people, expressed, in fact, a social demand in the form of a laconic time budget. Just about then, chronometric time-and-motion studies were introduced into industrial practice by the pioneers of "scientific management." These also meant time budgeting of a sort, by setting up precise accounts of the amounts of paid time spent by workers on all kinds of "necessary" or "wasteful" activities during their work in the factory.

The bulk of time-budget studies published before World War II originated in Great Britain, the Soviet Union, and the United States, with quite a few in France and Germany; other countries were only sporadically represented.[2]

In general, these earlier studies focused on the following topics:

a. the share that such broad categories of activity like paid work, housework, personal care, family tasks,

[2] For a more detailed review of the origins and the history of time-budget research see two essays by the author of this chapter: "Trends in Contemporary Time Budget Research," in *The Social Sciences: Problems and Orientations* (The Hague and Paris: Mouton/UNESCO, 1968; New York: Humanities Press, 1968), pp. 242-51; and "Trends in Comparative Time Budget Research," *The American Behavioral Scientist*, IX (May 1966), 3-8.

sleep, and recreation have in the daily, weekly, or yearly time budget of the population;

b. characteristic time expenditures of certain social groups or strata (e.g., industrial workers, farm homemakers, college students, unemployed men) on more or less specified types of everyday activities;

c. the use made of "free time," especially: leisure.

Data were collected mostly by means of simple forms on which the respondents were requested to report the time and the general character of their various doings during the day. Quite often only the remembered (or guessed) average duration or frequency of specified activities in the daily life of respondents was asked. Some moderate use was also made of "yesterday interviews" intended to reconstruct the course of events in the life of respondents on the preceding day. Finally, for some global accounting purposes certain estimates were secured by analysis of available data on traffic densities, cinema attendance, newspaper circulation, participation in social organizations or gatherings, and so forth.

It should be stressed, however, that this general characterization of topics and data-collection techniques prevalent in time-budget research before World War II does no justice to some exceptionally fine studies falling in the same period but foreshadowing much later developments. As early as 1924, for instance, quite advanced social statistical concepts were introduced by S. G. Strumilin in time-budget surveys done for the purposes of governmental and communal planning in the Soviet Union. (Strumilin's study was repeated thirty-five years later on a similar sample of the working population by his former pupil, G. A. Prudensky. So far as we know, this represents the first use of time budgets for purposes of historical comparison). In the early thirties the famous "Westchester Survey" of G. A. Lundberg and his associates opened up a whole new era of studies on leisure in the United States. Somewhat later, P. A. Sorokin and C. Q. Berger developed rather sophisticated conceptual approaches for the analysis of very detailed personal diaries covering relatively extended periods of individual daily activity; their famous book *Time Budgets of Human Behavior* demonstrated some surprising insights into psychological and social motivations starting from a careful analysis of time-budget data. In Great Britain, Kate Liepmann's excellent monograph *The Journey to Work,* published during World War II, concentrated on a problem which has since grown to gigantic proportions, namely, that of commuting and the length of commuting time. More examples could be given. It remains true, nevertheless, that everything that has been done in the field of time-budget research during this earlier period served only as a prelude to the momentous development which took place from 1945 onward.

This postwar development was characterized, first of all, by tremendous growth in the scope and number of time-budget studies. The main motivating force was presumably the general upsurge of scholarly and public interest in the study of social phenomena by means of mass observation, especially by polls and all sorts of sample surveys giving easily digestible, though perhaps not always as easily interpretable, quantitative results.

Many refinements of sampling, interviewing, registering, and coding techniques that were developed in various branches of survey research were quickly adapted to the needs of time-budget studies. In fact, from this period on, time-budget research has become, to a very great extent, a branch of survey research, and many institutions that specialized in the execution of surveys (statistical services, survey research centers, public opinion and marketing research firms, etc.) became deeply involved in it. This had along the way a beneficial effect, insofar as it pushed into the background some rather crude and inexact methods of data collection that were rather widely used in earlier time-budget research. On the other hand, it somehow seems to have hindered the adequate development of a field of great potential interest to psychological research: the analysis of personality traits expressing themselves in the individual's "treatment" of time, in his "time husbandry." We are, for instance, apt to characterize some people as notorious latecomers, others as being always in a hurry and arriving everywhere on the stroke; how a person behaves in temporal relations, how he estimates and remembers time, is probably very deeply rooted in his character. But we have very little exact knowledge about all this, although some aspects of the problems involved could very well be studied by the adaptation of some observational and analytic methods used in time-budget research to the psychological study of individuals in their daily transactions. However, studies of this kind are rare and lie, at least at present, far outside the actual boundaries of those studies on the use of time which we normally comprehend under the notion of time-budget research. In any event, the vast majority of contemporary time-budget research is based on group- or mass-observation techniques.

The application of modern survey research methods and of electronic data processing made it possible to develop time-budget research projects of a scope which nobody could have foreseen. Let us mention as an example the time-budget survey based on 170,000 systematically conducted, technically refined and well-controlled "yesterday interviews" carried out in 1960–1961 by the Japanese Radio and Television Culture Research Institute. It covered the daily activities of all strata of the Japanese people, in all parts of the country, in metropolitan, urban, and rural settings, during all the four seasons, on weekdays, Sundays, and special holidays. One of the main aims of this gigantic survey was to obtain reliable and very detailed data for planning and organizing radio and television programs, but many of its results could be used, and have been used, for many other purposes too. Indeed, this survey proved to be so useful that it was repeated in 1965, this time covering a nationwide sample of 24,000 mandays, so as to check, among other things, the influence of the spread of television on the daily habits of the Japanese people.

As another characteristic instance, we may put forward the fact that, due to the rapidly growing number of time-budget surveys carried out on behalf of governmental and communal authorities in the Soviet Union, a special conference of Soviet experts was convoked in 1960 so as to work out standardized instructions on the preparation of "time budgets of workers, technicians, engineers, and civil servants." Between 1959 and 1965 alone time-budget surveys involving well over 100,000 man-days of recorded human activities were carried out by the Institute for Economics of the Soviet Academy of Sciences in Novosibirsk and by some related research organizations, partly for general scholarly purposes, partly for supplying much-needed data to authorities concerned with the planning of manpower resources, educational facilities, communal services, etc. National statistical offices likewise became interested in time-budget research. In Hungary, for instance, where periodic micro-censuses (nationwide sample surveys of demographic changes in the population) have been introduced to bridge the long gaps between the regular census years, the micro-census of 1963 served also for obtaining time budgets of the population all over the country. The results have since been published by the Hungarian Central Statistical Office. This is probably the first instance where time-budget data were collected in a census context; the sample included 12,000 persons chosen from as many households.

Needless to say, by no means all time-budget surveys are currently based on population samples including many thousands of people. For the study of most problems no nationwide sampling is needed: survey sites are often restricted to a single region, a town and its surroundings, or a small settlement. In the majority of cases the survey is aimed at some specific groups or strata of the population, with a traditional preference for workers' families, employees, and housewives, and a more recent preference for youth, elderly and retired people, vacationers, and all sorts of other candidates for leisure.

Indeed, leisure, or the lack of it, is the central theme of an incredible number of time-budget studies carried out since World War II in practically all countries where social research has reached a certain stage of development. France, the Federal Republic of Germany, and the United States probably lead the race as far as the sheer number of time-budget studies focused on leisure is concerned, but some less affluent countries such as Poland or Hungary are good runners-up. In fact, the shortening of the daily or weekly working time and the lengthening of the average life beyond retirement age, the long hours spent in commuting or, for that matter, in sitting before the television set, the growing need for adult education, and the everlasting domestic slavery of housewives and mothers in spite of all refinements in frozen and precooked food, kitchen gadgetry, laundry automation, and the like, are all factors in studies of leisure. These factors have contributed to making leisure today a far more complicated and far more general problem than it was in a period when working people simply had to fight against impossibly long working hours that left them insufficient time for even their most immediate personal needs.

The leisure problem undoubtedly gave a great impetus to time-budget research. The fact, however, that time-budget studies that concentrated on leisure became so very fashionable and widespread had some less favorable consequences too. In quite a number of these studies, most nonleisurely activities figuring in the daily time budget received such superficial treatment, if any, that this considerably reduced the value of the whole analysis. It goes without saying that leisure can only be meaningfully interpreted within the total context of the activities of human life. By this we do not, of course, want to call into question the fact that very many time-budget studies of the highest distinction and of outstanding importance were done by researchers concerned with the investigation of leisure. Indeed, the first attempt at an international comparative field study on the use of "free time" was launched, under the aegis of UNESCO, by a group of specialists in the sociology of leisure, namely Joffre Dumazedier and his associates.

Time-budget surveys have also become important tools of urban sociology and have been applied to many problems of urban planning in several countries, especially in France and in the United States. As far as industrial sociology is concerned, new approaches to problems connected with schedules of work, rotating shifts, automation, etc., have been opened up by making use of time-budget research methods; Soviet, British, and French researchers have been very active in this respect. Applications of time-budget research techniques to operations research in industry have been specially developed in the United States, where these same techniques were also applied for the first time to the investigation of temporal relations in research work (time budgets of research laboratory staffs), a subject now enjoying some popularity among British, German, Polish, and Hungarian scholars interested in "research on research." A surprising number of various time-budget surveys have been carried out in some small countries whose research literature is little known internationally due to language barriers. Bulgaria might serve as an example: a professional research group attached to the Trade Union Council has carried out a series of rather extensive time-budget surveys for the study of labor problems.

All of these propitious developments provided advantageous conditions for the initiation of a research project which could provide the framework for the cooperation of social scientists in many countries who happen to have a common interest in this type of study and who are willing to undertake joint efforts to achieve some further, and better-integrated, progress in their chosen field of work.

Also, around the mid-sixties, general political conditions became much more favorable for promoting concrete collaboration in empirical social research between scholars in what may be referred to as Eastern and Western, or Socialist and Capitalist, countries. In earlier years any kind of actual research cooperation between social scientists of the East and West had been rather sporadic, and certainly no major comparative project involving extensive field surveys, full exchange of data, and common evaluation had been undertaken in cooperation between East and West.

The European Centre for Coordination of Research and Documentation in the Social Sciences was set up in 1963 in Vienna as a permanent exterior organ of the International Social Science Council with the purpose, among others, to further just this kind of research cooperation between East and West. One of the very first projects launched under its sponsorship was the Multinational Time-Budget Research Project.

As a matter of fact, time-budget research is one of the very few branches of empirical social research in which a somewhat comparable activity has been developed and a somewhat similar body of experience has been gathered by social scientists in the East and West during the lengthy time of their mutual isolation due to the conditions of the cold war. This proved to be rather helpful when Multinational Comparative Time-Budget Research Project undertook to bring together a relatively important number of researchers and research institutes of the East and West in a collective venture having the following aims:

1. To study and to compare in different societies

variations in the nature and temporal distribution of the daily activities of urban and suburban populations subjected in varying degrees to the influences and consequences of urbanization and industrialization;

2. To develop methods and standards for the collection and evaluation of data pertinent to temporal and other dimensions of everyday activity ("time-budget data" in the broader sense) which, apart from their interest to social theory, are also of considerable importance for the organization of working life and for the creation of satisfactory conditions for the enjoyment of leisure;

3. To establish a body of multinational survey data on characteristics of everyday life in urban surroundings under different socioeconomic and cultural conditions which could serve as the basis for testing various methods and hypotheses of cross-national comparative social research;

4. To promote, in general, cooperation, standardization of research techniques, and the exchange of quantitative data at an international level among social scientists involved in survey research who are endeavoring to achieve comparable results with a view to their common evaluation.

In spite of the great and widely acknowledged importance of cross-national comparative survey research for the general progress of social science, the number of such projects involving field work in more than a few countries has been rather restricted in the past. Many more were planned than could go effectively into the field, and many of those that went into the field could not be brought to a fruitful end.

Those that were successful achieved their aims mainly in one of two ways. Either partnership in the project was limited to a relatively small number of countries where well-equipped research organizations could be found, with adequate financial and manpower resources, having a common interest in the subject, and able to take a full share in the very considerable effort and expense such a project involves. Or, one powerful research organization worked out the general plan of a cross-national comparative research project and found the very considerable means necessary to send its own research teams or research fellows to a series of selected countries to carry out the local surveys, or to obtain the cooperation of local research groups which, if provided with the necessary financial and technical assistance, could carry out this work according to the provisions of the general plan.

Both these approaches have their drawbacks. The first tends to limit the cross-national comparative project to countries in which well-equipped research organizations, disposing of adequate financial and manpower resources and also having some common "culture" of survey research, exist. The second is not subject to this limitation but tends to allot a minor role in the development of the basic research design and also in the comparative evaluation of the cross-national data to the participating local research groups, even if the best of efforts is made to take account of their views as fully as possible.

A truly multinational organization of a cross-national comparative research project presupposes that the basic research design will be developed collectively by all participants in the project, all standards of survey procedure (choice of survey sites, sampling, interviewing, coding) will be worked out and agreed upon in common, and the whole body of collected data will be made accessible to all for evaluation and interpretation, giving ample opportunity, on the one hand, for the elaboration of common views and ideas of the participants about the results achieved and, on the other hand, for the validation of their individual approaches and judgments.

Admittedly, this is an arduous way to carry out a project. It has, however, obvious advantages in making better use of the collective insight and wisdom of a greater number of social scientists with a greater variety of personal background and experience, and also is promoting better understanding and a fuller measure of cooperation between social researchers in various countries. It lays also the groundwork for future, better-founded enterprises in the field of cross-national comparative social research by helping researchers in countries where survey research has not yet been fully developed to obtain assistance and enrich their experience through participation in an international research project of this kind.

The level of technical preparedness and methodological sophistication in survey research is very different in various countries, especially in such diverse and widely dispersed countries as those participating in this project. Moreover, every research team had to carry out the survey it undertook by using the domestic resources it could mobilize. Naturally this necessitated some modesty and compromise in the general research design and in the scope and methods of data collection. In a centrally planned, directed, and financed cross-national comparative research project perhaps more ambitious standards might be set up and more uniformity achieved.

Agreement also had to be reached between all participants—again resulting in some compromise—about the categorization of everyday activities for the purpose of registration and coding. Political, cultural, and ideological differences play a great role in determining people's views about which activities in everyday life are the more relevant and should therefore be analyzed in greater detail, and about how activities should be classified or even merely named. Perhaps sometime in the future, sophisticated methods of computerized content analysis will make it possible to analyze vast numbers of detailed personal reports of people on their daily activities in terms of the "own words" of the respondents, but presumably even then some scheme will have to be developed for identifying and classifying those expressions. At present, however, a very elaborate nomenclature and classification covering all possible types of activity has to be conceived and worked out in advance for the registration and coding of respondents' daily doings. Moreover, this nomenclature and classification has to be constructed so that it is (a) detailed enough for all the various purposes of analysis to which participants may wish to subject, collectively or individually, their own and their partners' data; (b) not so minutely detailed as to overburden the interviewers, coders, and data-processing facilities available to the participants; and (c) clear and unambiguous enough to be translatable into a number of different languages and applicable somewhat uniformly in widely varying social and cultural settings. In this respect again, a highly centralized research organization, especially one possessing an itinerant research team to direct or supervise the work done at all the various survey sites and to carry out all or most of the analysis on its own, may be in a more advantageous position.

Although difficulties similar to those encountered in connection with the establishment of unified nomenclature and classification of activities had to be overcome with respect to many other types of data as well, an effort has been made to avoid any oversimplification of the research design or to impose any excessive limitation on the dimensions of the observation. Indeed, the daily doings of the respondents were recorded individually, taking account of what (and what else simultaneously) they did during the day reported, for how long, how often, at what time, in what order, where, and with whom. So as to ensure full freedom of movement in this multidimensional property-space for future analyses, every single episode reported in the 24-hour "life history" of a respondent was registered and subsequently coded in a way permitting its retrieval and reclassification according to any combination of its coded characteristics. This preservation of elementary data in their original context, and in an unaggregated form of course, put a practical limit to the number of dimensions recorded and the minuteness of registering and coding observations. Still, we think the array of data on various temporal, locational, and social attributes of daily activities collected within the framework of the Multinational Comparative Time-Budget Research is somewhat unique in the field of study, especially if one also takes into account the rather comprehensive data on personal, family, household, residence, and working conditions, etc., characteristics of the urban and suburban populations sampled.

JOFFRE DUMAZEDIER

LEISURE AND POST-INDUSTRIAL SOCIETIES

What is the meaning
of leisure for the masses of post-industrial society?
What will be the future meaning for these same
groups? In relation to this new phenomenon of leisure,
what is and what will be the meaning of work, family
obligations, religion, and political responsibilities?
More generally, in the transitional period from industrial to post-industrial society, what happens to the
relations between society and the individual? For the
last ten or twenty years this central question has been
present in all sociological thinking on leisure from
David Riesman or Harold Wilensky of the United
States to Radovan Richta and Blanka Filipsova of
Czechoslovakia. To analyze this new phenomenon within a relevant framework, we need a new theory, yet
we have only an old one at our disposal. Still, it is too

early to build a new theory. Before reaching this stage, however, we need to do a dimensional and critical analysis of leisure within different economic, social, cultural, and political contexts of advanced industrial societies.

To progress in this analysis, we require further investigations using comparative and projective methods. At the beginning of industrial society it was necessary to wait a century, from Adam Smith to Karl Marx, to understand the meaning of the change in work and the corresponding change of social structure in the dynamics of that new society. Very few countries in the world are at the beginning of post-industrial society. What can ten years of scientific effort to collect data and to try to integrate these data into a relevant framework tell us? The history of science gives us a lesson in patience. So let us admit that today we are not able to present a theory. It would probably be only an ideology in the sense of an anachronistic framework that would not lead to an understanding of the new reality. At the beginning of the first comparative study on the socio-cultural meaning of leisure in different social strata in seven countries, which are at an intermediate stage between industrial and post-industrial civilization, we can only state two assumptions. These we are trying to transform into verifiable hypotheses in order to reach a better understanding of the sociological meaning of leisure "in the post-industrial stage of this technologically oriented civilization."

— 1 —

Many sociologists limit their observation on the evolution of leisure to twenty or thirty years. They find that many workers today don't have more free-time activities or other activities than they had before. But it seems to me that a better way of studying the socio-cultural mutation of leisure in a new society is to compare the new state of leisure in the worker's new way of life in the post-industrial society of today and tomorrow with the state of leisure of the entire industrial society since its beginning. If we don't do that, we risk attempting to solve a structural problem of a new society with conjunctural information concerning a short period of a past society. This method would be irrelevant. Only one method—combining the historical and projective—is, in the long run, pertinent to solving this crucial problem.

In preindustrial societies, most of the workers were agrarians. In European countries, France, for instance, they worked all day until sunset. Each week they respected the day of the Lord. Each year they were kept idle for about 160 days by diseases, bad weather, and religious holidays. In an industrial society, industrial work becomes the most important factor in the dynamics of change. In this society, little by little, industrial workers become a majority and the largest part of the population becomes urban. At the beginning of this society the worker works ten hours daily—almost as much as in rural life—but under terrible working and living conditions. The week does not always end with Sunday, but often with an extra day of work. And the year is only a year of work, with many days of illness or "wine-bibbing," few holidays, and no paid vacations. Increasingly the technological progress of production, the appearance of the new work values, and the struggle of trade unions transform the situation. The struggle for better wages leads to the demand for more free time. Out of this context the philosophy of work for post-industrial society has been elaborated. Productivity from industrial work remains the most important factor for the economic development of the entire society, but the quantitative progress of free time has brought about qualitative changes.

The majority of workers no longer belongs to the industrial sector, but to the sector of services: distribution, administration, education, culture, leisure, etc. Work time has been reduced to less than 40 hours within a five-day week. The amount of free time for a week now tends to be equal to the duration of professional working time. For instance, in Jackson, Michigan, working time for a man without children is about 39 hours and free time about 35 hours, with about one hundred and forty days free of professional work each year. By the turn of the century, about thirty years from now, Kahn and Wiener foresee, with the development of atomic and electronic devices, a possible reduction of the working day to $7\frac{1}{2}$ hours with a four-day weekend, for instance, Friday, Saturday, Sunday, and Monday, and a thirteen-week vacation—like teachers.[1] In the advanced industrial society of the USSR, sociologist B. Gruschin agrees with predictions of the Russian Communist Party that before the end of the century weekly free time of 45 hours will probably be much longer than weekly working time. The predictions of physicists such as Kahn or of economists need to be based on projective and comparative research of all the social sciences working together. Many factors, not only technological and economic, but also

[1] Herman Kahn and Anthony J. Wiener, *The Year 2000* (New York: The Macmillan Co., 1967).

social and cultural are involved. To a sociologist, such anticipations as those of Kahn seem too linear and also too optimistic. But to guard against ideological and anachronistic thinking on leisure, it is necessary to remember that during the evolution of European or American advanced industrial society a year's working time has decreased from about 3,500 hours to about 2,000 over a period of 100 years, according to Jean Fourastie. What will happen thirty years from now if the past tendencies continue, or if they increase? In our opinion, it is not the evolution of the length of free time, past or future, which is most important in itself. What is most important is the mutation of cultural values brought about by this modification in the use of time in mass society. Our central hypothesis is that the most important effect of the change in the time structure of the workers' new life is a fundamental reconsideration of the relation between the values of individuals and the values of society, between the rights of individuals and their social obligations, all of which were a product of industrial societies. This is the origin of a slow but more and more determinant cultural revolution.[2]

— 2 —

It is difficult to know exactly if, for the majority of workers, the value of work tends to diminish with the shortening of the duration of working time. But according to the historical studies in the sociology of work this is likely. Although industrial work remains the most important economic factor of the basic dynamics of post-industrial society, the majority of work-

[2] John E. Goldthorpe, *The Affluent Worker* (Cambridge: Cambridge University Press, 1968).

ers and employees (blue collar or white collar) in American society do not sense fulfillment of their personality in their work (62 percent of blue collars and 61 percent of white collars).[3] Indeed, Daniel Bell said that in the post-industrial enterprises we are witnessing the rise of a new type of technician. However, this new group (now 10 percent in the United States) will be in the minority for a long time. The key question remains, What happens or what will happen to the majority of employees when work becomes more specialized and less interesting? The same observations hold true for the USSR and Czechoslovakia, based on empirical studies of cybernated factories, to wit: a small number of jobs require more responsibility and education; however, if the objectives of rational production remain, and even if participation by workers through their organizations is encouraged, the situation remains that for the majority of employees self-fulfillment through work is doubtful.

We now come to the main problem of this chapter; what happens to the time freed from occupational work and to corresponding values? Time liberated from work is not necessarily leisure. It includes time for domestic and family work. According to the American, Nels Anderson, time liberated from work has been devoted to family activities or private activities. This is true, but what are family or private activities? It is a confusing concept based on views of the family that don't correspond to the new family group with its involvements in other groups, its new activities as a group, and the activities of its members as individuals, whether "other-directed" or independent, especially

---

[3] C. Wright Mills, *White Collar: The American Middle Classes* (New York: Oxford University Press, 1951).

among young people during their leisure time or semi-leisure time. It is necessary to distinguish between family obligations (domestic work and educational duties), semi-leisure activities (do-it-yourself projects, etc.), and leisure activities with or without the family group, whose main purpose is individual satisfaction.

Indeed, we would suggest making a special study of married women who work. The percentage ranges from 30 percent in France or the USA to 85 percent in the USSR or Czechoslovakia. For these women, time off from their occupations is taken up with family duties much more so than for the working man. Szalai's Jackson, Michigan survey revealed that, in an average household, the amount of free time per week for a man is 32.6 hours, while a working woman, after she has done the household chores, has only 29.2 hours. Even when there are no children, there is also an inequality: 35.6 hours for the man, 33.9 hours for the woman. Variation of the economic and political system doesn't change this cultural inequality. In fact, we find a more pronounced inequality between the sexes in the industrial society of the USSR. Among urban common laborers the working man has 3.39 free-time hours daily, the working woman 2.42 hours. This may be due to fewer home appliances and a lack of transportation. Perhaps the particular cultural patterns can help explain the differences. What remains, however, is a long-range trend toward reduction of time for domestic work, even for working women.

This trend has been proved by several American authors, such as Nelson Foote, David Riesman, etc. In the same perspective, Strumilin observes that in the USSR the time allowed for domestic and family duties among urban workers, in 1959, was 1.72 hours per day for men and 4.80 for women; in 1924, 1.70 for men, 3.91 for women. For the family as a whole, the greatest increases occur in semi-leisure or leisure activities such as holiday trips, weekend or evening outings, television shows, listening to music, participation in clubs, etc. These activities are family affairs, under the same roof, in the same car, with or without friends.

We observe the mutation of values that accompanies the decrease in family duties. This mutation comes about through leisure activities (television, movies, trips, games, etc.), creating a family that is other-directed as compared to the inner-directed family of the industrial society. Among the younger generation the discovery of artificial contraceptive devices has transferred the most important part of sexual relations from the realm of the reproduction or risk of reproduction of the species to the realm of games between individuals: husband and wife, lovers, or friends. This creates a new problem as to the ethics of the game. Indeed, the Kinsey Report had a bomb-like effect upon American society when it revealed in 1953 that 90 percent of American men and women were engaged in sexual games that were against the rules of their religion.

We should also mention here what is called "juvenile delinquency," an expression that is juridically clear but sociologically obscure. What part of delinquent behavior is a genuine breaking of the rules of the new society, and what part of delinquency is merely a breaking of out-of-date juridical laws that applied to the former type of family and society? Indeed, family cohesion is not threatened with the new society, but renewed—thanks to leisure and semi-leisure relationships and activities.[4] Last but not least, the new family group has a tendency to open up to the values of the new society, especially to individualistic ones.

In time freed from occupational work and family obligations, what happens to religion or spiritual activities? Regular religious practice, among all social classes, has decreased in Europe and the United States since the end of traditional society. In France, for example, for the last twenty years even holidays have been experienced by the majority as leisure activities without family, civic, or spiritual connotation, except Christmas. For the majority, Sunday is no longer the Lord's day, but a day added to the weekend in which fishing and hunting are more popular than praying. On the other hand, the minority who are devout Christians are probably more active than yesterday. For the majority the weekend comes first, and this new society might be called the "barbecue society." In terms of values, holidays tend to become, especially among young people, artists, and intellectuals, not a time for social, institutional ceremonies, but a time for brightening daily life through exceptional communication between individuals or groups, a kind of "quasi-sacred" climax for love or friendship in social leisure, halfway between routines associated with institutional obligations and the dream world where the individual has a maximum illusion of liberty or fantasy.

Following this brief look at religious obligations, it is interesting to wonder what part of free time is invested in political activities. It was the explicit hope of Friedrich Engels that a reduction in the time spent

[4] Erwin K. Scheuch, "Family cohesion in leisure time," *Sociological Review*, VIII, July 1960.

in work would permit a larger and more intensive participation of workers in politics. What is the actual situation in this respect? Minorities seems to be using a great part of their free time to enter into political life or participate in political activities.

However, in French society, electoral abstention and nonparticipation in unions have remained practically constant in spite of changes in the structure of working time, and in spite of drastic changes in the political regime itself. In France, for example, except for a few short periods of high participation, the rate of voluntary participation in unions has remained around 80 percent. In most advanced, capitalist societies, permanent participation of political parties appears to be declining. Of course, besides parties and unions, we have witnessed the development of a great many voluntary associations for the encouragement of social and cultural participation. Of course some of these associations are permanent or temporary pressure groups representing a wide range of political positions. But the significance of most associations in terms of politics at large is limited or ambiguous, and their development seems to be jeopardized in urbanized societies.[5] One would be tempted to suspect this decline only in advanced industrial capitalist societies. But the studies made by Strumilin and Prudensky on the evolution of the use of free time by the Russian urban worker are significant. Between 1924 and 1959, participation "in all social activities," including political and union activities, decreased from 109 hours a year to 17 hours a year for urban manual workers, in spite

[5] Murray Hausknecht, *The Joiners* (New York: The Bedminster Press, 1962).

of the increase in free time. The same development is observed by sociologists in Czechoslovakia. In view of these facts, the team of Marxist philosophers, sociologists, and economists led by Richta concludes that "if new forms for participation are not developed in the near future there will be a serious political vacuum even under socialist conditions."

Another activity that has greatly benefited from the increase of free time in the advanced industrial stage is education. Educational activity has replaced the vocational activities which operated at the beginning of industrial society. We refer here to educational activity in secondary schools and in universities. Post-industrial society seems to have produced, or is producing an academic revolution for a fast-growing number of people, according to David Riesman. But these are activities imposed by society or by the family on the young people. And the latter are not always satisfied when they have to fulfill these new obligations.

The social movements of students constitute a revolt against the schooling system, especially in advanced industrial society: Berkeley, New York, Berlin, Prague, Paris. The achievements of our educational system are being criticized. A comparative study in the USA, Great Britain, and Sweden between university achievements and achievements accomplished in real life has shown no positive correlation.[6]

Steiner's national survey has also shown that less than 10 percent of the people who have attended grammar school have a TV preference for high-level programs of a scientific or artistic nature. But the per-

[6] S. H. Halsey, Jean Floud, and Arnold C. Anderson, *Education, Economy, and Society: A Reader in the Sociology of Education* (New York: The Free Press, 1961).

centage is also below 10 percent for high school graduates, and it is only 11 percent for college graduates. On the other hand, in post-industrial society the voluntary education of adults is increasing rapidly. About 20 percent of the American population, or 25,000,000 people, participated in 1965, which is four times more than twenty years before. Voluntary education is, for the majority, taken on as a leisure activity. In face of the dubious results of imposed educational schools and universities and in a society where technical and cultural models change faster than do generations, it would be perhaps useful not to prolong compulsory schooling for the young people, but to invent a new system of permanent education for both youth and adults that is oriented toward the self-development of the personality at every age, creating a new relation between obligation and free choice, between study and leisure.

We can conclude that the main uses of time freed from work in the beginning of post-industrial society probably are leisure-related. In the United States, Szalai reported, leisure activities occupy an average of from 3 to 5 hours daily for the majority of workers. For most workers more than 140 days yearly are free from work. Leisure is not made up of isolated activities. But it is a new cultural system with interrelations among activities, interests, representations, values. Leisure has temporal dimensions: leisure at the end of the working day, at the end of the working week (weekend), at the end of the working year (holidays), at the end of the working life (retirement or third age). An interrelation exists among these four periods of leisure. One cannot seriously juxtapose leisure and culture. Leisure is the temporal framework for the cultural content. This content may be low, medium, or high "brow" depending on the standards, capacities, knowledge, and values of the creators. Within the limits of the social determinants of each milieu, the individual chooses how he will use this leisure and what its content will be.

This is a new social right of man. Old social duties, work, family obligations, political or religious activities, are to be reconsidered from the standpoint of these new values and norms. We do not automatically find here a regression of institutional obligations, but we find a need for their transformation according to the new cultural system of leisure. This trend becomes central in the new aspirations of individuals, especially among young people and part of the intelligentsia. This aspiration tends to be dominant in each leisure activity: physical (sports), practical (handcraft), artistic (creation or contemplation), intellectual (reading magazines or books), or social (going to a dance or attending an association meeting). Mass culture is part of the content of mass leisure. Finally, leisure brings together entertainment and learning, rest and free creativity, thanks to its psycho-sociological dimensions. Leisure activities liberate the individual's personal forces from the fatigue that results from work and other institutional obligations by rebuilding these physical or psychical forces. Leisure activities allow the individual to keep his personal energy and allow him to escape boredom through entertainment. Finally, it allows the individual to perfect his personal resources, so to speak, to invest in himself.

In the Renaissance the individual achieved the right to choose his own God or his own ideal without risking being burnt at the stake. With the coming of democratic society in the eighteenth century, the individual achieved civil rights, i.e., protection from arbitrary will of political power (habeas corpus). The trade-union movement was a struggle against the arbitrary will of owners and management. The worker ultimately achieved the right to organize. All these rights have been achieved over the last four centuries. This conquest has reached a point of no return. In the new society the fact of leisure corresponds to a new social right for the individual, i.e., the right to have for his own use an increasing part of free time, leisure time. Society used to say, "To want this is to be lazy, idle, selfish." Today, we just say "leisure." All theories based on a conception of work, family duties, political activities, or educational activities are bound to fail if they consider leisure merely a compensation or a complement to institutional obligations needed for the functioning of economy or society, without reference to the mutation of the new individual's needs.

These new individual needs appear sooner or later in all social classes of post-industrial society. They are a new form of individualism, whose progress has been analyzed by Riesman, at the very moment that Americans were threatened by social conformism and "groupism." Even in socialist countries more theoretically oriented toward the exaltation of collective values, the same revolution appears when the productive forces near the post-industrial level. "The human development of the individual," said Richta, "should be honored as a collective value." This new social right for the individual takes its place in the long struggle for basic human rights.

There are two objections to this. This framework is not relevant for people from all nations and for all social milieus. What happens to people who don't have time, or who have only very little time, for leisure? Of course, it would be a delusion to believe that leisure will be equal for all and that mass leisure will do away with class leisure. Even in American society, striking ethnic and social differences in leisure activities, and in the kinds and levels of cultural interests related to leisure, have been revealed by empirical studies made by Wilensky, Steiner, and others. In the field of leisure, too, there is still another America, according to Harrington's expression: an America of socially and culturally unprivileged people (20 to 25 percent).

These are exceedingly important social problems. But sociology does not confine itself to the analysis of social problems of the minority. If sociology focuses on problems of the minority instead of studying the problems of the majority, it runs the risk of losing the sociological imagination that is necessary for understanding the new society. Including both old and new social problems, there exists a form of "social wretchedness," which does not just look at social differences, but investigates the helpless unawareness of sociological change of the majority of people and, in the practical field, the inability to invent a better mass society guided by human values.

What about other inequalities that are even more important, such as those existing between the high level of production of services or free time in post-industrial societies in North America and the low level of production of goods, and the high level of unoccupied time in pre-industrial society, as in South America? Unoccupied time for the masses in South America is the contrary of free time in North America. In the former, economic development requires more hours of work; the latter requires more hours of free time. One of the most important social problems for America is, perhaps, to harmonize through an economic and cultural policy the increase of leisure time for the 200 million inhabitants of the rich, North American society and the increase of working time for the 300 million inhabitants of the poor nations of South America. This harmonization is especially linked to the economic and cultural exchanges between two civilizations. To increase our knowledge of the conditions for this harmonization, we need comparative studies that would test propositions based on the development of work in pre-industrial societies, and not those coming from the problems of increasing mass leisure in post-industrial societies.

For post-industrial society the work ideologies, community life, and social participation that grew out of industrial or pre-industrial societies will not suffice as a solution to the national or international problem of solidarity. We need to invent a new mentality, a new conception of economic and cultural exchange through work or tourism. If a cultural revolution is necessary, it will be a cultural revolution adapted to post-industrial society, by people of post-industrial society. In our opinion, to deny the new needs of the personality, which appear with the production of leisure by post-industrial society, is a fantasy. To understand the true possibilities of these new aspirations of the individual aspect of leisure, we need a new economic and cultural framework. It is too early to build a true theory of this new phenomenon. First, it is necessary to eliminate from the field the communitarian, puritan, or totalitarian ideologies, which date back to a time when the problem of mass leisure of post-industrial society and its new values arose.

This is not the first and will not be the last time that the growth of a scientific field started with a fight for liberation. It is against the idealistic illusion that Adam Smith and Karl Marx founded a science of work for industrial society. It is against the psychological illusion that Emile Durkheim created the sociological field itself. Sociology of leisure began with the discovery of the field of a new social right of the individual against the anachronistic obligations of work, family, religion, and politics of the passing society.

— 3 —

How can we analyze the dynamics of leisure in post-industrial society? To answer this question we need to formulate another question, What are the main dynamics of post-industrial society? For the last twenty years, many sociologists have been studying the characteristics of post-industrial society, focusing on the new scientific revolution, the new technology of cybernetics and automation, an economy of services, mass society, mass production, mass consumption, mass leisure, mass culture, mass education, and so on.

In 1967 Daniel Bell saw three dimensions to post-industrial society, of which the third is the most important:

—a shift from food to services;

—the rise of a large scale professional and technical class (14 million by 1980);

—theoretical knowledge essential as the source of innovation and policy analysis in society.

For us, this definition is too technocratic. What happens in post-industrial society to social relations and human values associated to work and nonwork? Radovan Richta shows the necessity in the new society to go beyond a work-oriented society to one which encourages the creative activities of individuals both in their work and outside their work, thanks to the scientific-technical revolution. This recent analysis by a Marxist philosopher is of particular interest. This is a new tendency of Marxism in which Marx appears more as a critic than an apostle of work.

David Riesman saw post-industrial society as one where maintenance of life is no longer a problem for the majority, where work and leisure even to a greater extent become of utmost significance. Where investment in a post-secondary education becomes as important as investment in capital and where relationships between men and women, between old people and young people become even more fluid than in the past.[7] These characteristics of post-industrial society are interesting, but what is the fundamental dynamic creating and developing this society?

Our opinion is that we need:

a. to reanalyze the fundamental quantitative process of industrial society in which there is an increase in production and a decrease in the amount of work.

b. to make a new type of analysis of the qualitative changes, which are the consequence of this quantitative process of economic, social, cultural, and individual life, whereby industrial society becomes post-industrial.

A new scientific discovery transformed the power of man over nature. Now we can use machines not only

[7] See David Riesman, *Abundance for What?* (New York: Doubleday, 1964), pp. 162-83.

to take the place of the body in work, but also to increase the power of the mind to reduce the uncertainty of the future. For the first time, human action itself becomes a scientific problem through the discovery of linear programming, the analogue models of action, and the simulation of possible alternatives for decisions. This new scientific power gives hope for the transformation of the human praxis into language.

A higher status is given to innovation, and a lower status to traditional knowledge and routine experience. This increasing status of innovation tends to spread from scientific knowledge to the practical, artistic, and ethical fields. Innovation tends to diffuse from specialists and invade one's life. This movement makes old conflicts more acute or produces new conflicts between innovators and traditionalists, between new and old generations. The crisis of information and education is growing more acute all the time. What knowledge should we spread? What learning should we acquire? For whom? By whom? In what time? With what method and within what structures? What culture does this new society need? Both an educational and a cultural revolution are beginning.

This new source of scientific knowledge brings about the development of a new system of mass production, thanks to cybernetics. At the same time, the system without cybernetics is often condemned. A minority of workers is becoming increasingly skilled, but the majority is becoming progressively less qualified and alienated. The increase in productivity allows productive forces to be diverted from agricultural and industrial work in favor of material and cultural services. The question of the correspondence between these services and the human needs of the personality has to be formulated on a mass-society scale.

This new kind of production transforms the active working population: the majority does not work any longer in the agricultural and industrial sectors, but in the service sector. A minority of the white-collar group becomes an increasingly rich and powerful group of technocrats, as Daniel Bell noted, but another minority without specialization or additional education becomes relatively more isolated. Social conflicts are not restricted to those among the old social classes, but appear through new social divisions among different groups of workers, among different ethnic or religious groups, and among different generations, or they appear as a result of resentments against the new groups of technocrats, bureaucrats, intellectuals, and so on. This active population is governed in its work and private life by mass organization which increasingly stimulates periodic or permanent rebellion of minorities who condemn the system itself.

Less work, for what? Time spent on family obligations is more mechanized and organized. It remains longer for the working woman, but is on a par among women, men, and young people. Solving this problem requires new values and norms in the realm of family obligations, so that the possibilities of leisure can be equalized among all members of the family.

This bulk time freed from work and family obligations is partly spent in religious and political activities. At present, the greatest part of freed time is dedicated to study by the young and to leisure by adults. However, in the new mobile society the time for study tends to continue throughout a whole lifetime. Therefore, the new society needs to do away with the present schooling system and build a permanent system of mass education for young and old. An increasing part of education comes during a worker's leisure time. The new society needs to reconsider the relation between study time and leisure time. This new system of permanent education needs to revaluate the relations between the values of Homo sapiens and those of Homo ludens.

Finally, in spite of economic difficulties and cultural lag, for the majority of workers, freed time has been, and will increasingly be, spent in leisure with its particular temporal, cultural, psycho-sociological dimensions. The growth of the new leisure values leads to new normative patterns for work, the family, religious life, and politics. These patterns are very different from the norms of industrial society. Now, what will be the cultural content of mass leisure?

Speaking normatively, this cultural content should stimulate an increasing participation of the majority in the liberation of underprivileged minorities of rich countries, and to the development of Third World countries. What values will stimulate the fight for justice, the search for truth, or the contemplation of beauty in the daily life of masses, instead of a passive consumption of goods or services, increasing for more and more people? This is the most important question which the dynamics of post-industrial society raises. In other words: what qualitative changes in the economic, social, and cultural life of individuals have occurred as consequences of the quantitative revolution of industrial society, namely, the capacity to produce more while working less?

Hence, in this perspective, *leisure is a part of time made free by the growth of the productive forces, and is oriented toward rest, enjoyment, and improvement of the individual, as an end in itself.*

— 4 —

How can we study the social and cultural dynamics of leisure in the evolution of post-industrial societies? When I was called upon to do research with economists on the probable evolution of French society (within the framework of Group 1985), I felt the need for a scientific and objective knowledge of the evolution of leisure and mass culture in the United States. In my opinion there is nothing so ridiculous as a certain type of European intellectual who is hostile to American patterns and, at the same time, totally unable to foresee the actual invasion of these American patterns; hence, there is an incapacity to prepare for it. The American challenge is not only economic, it is also cultural.

It doesn't seem possible for us to attempt a previsional sociology of leisure in the evolution of post-industrial societies without looking at American society. Such a choice doesn't imply an a priori preference for a given economic system. It doesn't imply either a positive or negative estimation of American society. We are not making a political or ethical choice. This is a scientific choice.

A comparative study is a preliminary obligation for any sociologist of industrial societies, whether capitalist or socialist, who wants to attempt to establish a sociology of post-industrial societies. In the same manner, the choice of studying English society was a scientific must for Adam Smith and Karl Marx in the nineteenth century. They sought to have a better understanding of the problems related to industrialization and capitalism. This helps to understand why Richta's scientific team in Prague and Kahn's scientific team in New York have reached the same position on this point, as it clearly appears from a comparison of the books they have published on the post-industrial American experience. Today American sociologists are alone facing the future. This fact is often forgotten abroad, because Europeans often have an emotional reaction to America—either admiration or hostility. Only after a private conversation with David Riesman did I come to a full understanding of a sentence he wrote in his book entitled, *Abundance for What?*: "We must invent our own pattern for the future as we move forward into a new historical situation."

For the sociologist, the analysis of American society is full of dangers. In *The Lonely Crowd* Riesman foresees that the social and cultural mutations taking place in the post-industrial society of the USA will reproduce themselves in all other post-industrial societies of the future. Is this likely to happen?

In other words, will the function and structure of patterns of leisure and mass culture be similar everywhere to those of the post-industrial American society? Wouldn't other patterns be possible, even in American society? Can the interaction of these post-industrial patterns with other national traditions, with other means of appropriation of the means of production and diffusion be considered possible?

These are questions which an analysis of the American situation alone doesn't seem to be able to answer. To find an answer, it is necessary to compare, in the light of previsional hypotheses, the evolution of the precise characteristics of leisure on both macro and micro levels and see this in relationship to the evolution of fundamental characteristics of society. This must be done not only with American society, but with others which are in the process of reaching the post-industrial stage or which are near this process.

In this perspective, one society is of great strategic interest although it is usually neglected by American sociologists facing post-industrial problems. I am referring here to Quebec, French Canada. Since the years 1955-60, Quebec has been awakening. First, led by a conservative ideology, and then by a "catching-up" ideology, Quebec is increasingly being carried away by collective plans or dreams of autonomous development. In spite of its economic dependence upon the United States, in spite of its problems of biculturalism, in spite of its economic backwardness as compared to Ontario, and in spite of its internal inequalities (Gaspésia-St. Jerome, etc.), Quebec society is already facing some problems of post-industrial society, reflecting the United States and English Canada.

The structure of its active population and the level of its production are already of a post-industrial type: its income per capita is the third highest in the world. In 1967, 8.4 percent of its workers were in the agricultural sector, 30.3 percent in the industrial sector, and 61.3 percent—that is to say, the majority—in services, as against 40 percent in France, for example.

By observing Quebec, it is possible to learn what becomes of post-industrial characteristics in a French cultural context, discounting as far as possible the Anglo-Saxon influence from the United States. What helps in discounting this influence is the fact that today the majority of the elite in Quebec and Montreal have stepped up their struggle for economic, cultural, and linguistic autonomy of their society.

At this point, we are up against another problem.

How can we eliminate from our field American variables, either Anglo-Saxon or French? Do post-industrial characteristics still appear once these variables are eliminated? To answer these questions, we must observe what happens to these phenomena in another context, a European context, for instance. In this perspective, we feel it is particularly useful to study two advanced industrial societies: Sweden and Switzerland (French- and German-speaking). We have added two countries to these: West Germany and France. Germany, because it is the most advanced industrial society in the European common market, with 11 percent of workers in agriculture, 48 percent in industry, and 41 percent in services in 1964; France, because its economic development closely follows the German development in a different cultural context, 21 percent of workers in agriculture, 39 percent in industry, and 40 percent in services in 1963.

In these societies, post-industrial characteristics are in full development, but this time in European cultural contexts. Are these characteristics similar today to those of the two Americas, the United States and Quebec? Are they going to be so in the future?

Finally, one can wonder whether the characteristics common to post-industrial society in these different countries are not the result of an economic and social structure that is dominated by capitalist initiative in production and distribution. However, it has seemed useful for us to observe, since 1960, the new tendencies of a socialist industrial society which is most advanced, thanks to its level of industrialization, urbanization, education, popular culture: the society of Czechoslovakia. The first social and cultural mutations of a post-industrial economic character have appeared or have started to appear in Czechoslovakia. Indeed, Russian society has had the means of developing some sectors of industry—aeronautics, for instance—which are particularly efficient and striking, but, if you limit your analysis to general characteristics of economic, social, and cultural evolution (taking into account the artificial delay imposed on the productive forces of the Czechoslovakian society for the last twenty years by the USSR through the discipline of the common market), it is obvious that the socialist society closest to the post-industrial stage is indeed that of Czechoslovakia. If you use two criteria, the levels of production and the structure of the working population, you can see that, in spite of the economic crisis Czechoslovakia is going through, and in spite of the iron collar imposed upon the country, it is capable of the highest production per capita among all the socialist countries. In Czechoslovakia, for every one hundred workers, 22 percent work in agriculture, as against more than 30 percent in the USSR, 46 percent work in industry, and, last but not least, 32 percent work in services, which represents the highest percentage for the sector of services among industrial, socialist societies, except Eastern Germany.[8]

If you emphasize the social and cultural characteristics appearing in the new society, the present "renovation of Marxist thoughts"—made necessary by the analysis of these new phenomena—is of decisive importance in our opinion, not only for the adaptation of socialism to the post-industrial characteristics of a society, but also for a better knowledge of the charac-

[8] For the methodological approach see Joffre Dumazedier and Marc LaPlante, "Projection and the comparative method," *Cahiers Internationaux de Sociologie,* 1969, pp. 69-92.

teristics of post-industrial societies themselves. It is necessary to supplement and correct the study of American society by applying the method of concomitant variations to these different contexts if we want to improve our knowledge of the specificity of post-industrial society, and if we want to have a better understanding of its different, possible types.

How can we make a comparative study of these different societal types to see whether there are common characteristics? We must look for:

a. the permanence of variables: for instance, cultural and social contents of mass leisure and their implications in other activities, such as work, family activities, etc.

b. The correlation of these contents with the specific characteristics of post-industrial societies, in spite of the difference of traditional cultures or the economical and political systems.

— 5 —

In our double hypothesis on the dynamics of leisure in the dynamics of industrial or post-industrial society we are faced with this central problem: More leisure for what cultural needs of mass society? To answer this problem requires imagination.

First, if productive forces offer new possibilities to free time while allowing the continuation of economic development, then what will this time be used for? Who will have this free time? How *should* this time be spent?

Kahn's forecast about leisure in the year 2000 appears too isolated from other possibilities for using free time according to different criteria of social and cultural development of mass society. The rise of free time for leisure is not automatic. The use of freed time should be the result of a policy just as is the utilization of income among social groups in each stage of economic development. What is better for the individual or the society for each social milieu? This is a policy question. In order to solve it there is a need for political discussions by experts and the population in general.

Some of the main alternatives, as suggested by John W. Johnstone in *Volunteers for Learning,* would be: Increase leisure time for all workers; promote time off for voluntary social and political activity to help in the creation and organization of community; provide time-off for those who would like to get further education in order to advance professionally (actually about 30 percent of adult education) as well as for other reasons which may be less instrumental; increase the length of time for the education of young people. If this is done, what would be the aim of prolonging education? and for whom? Perhaps it would be well to reduce the working time of active women by one or two hours a day, so they could devote that time to household duties or education without sacrificing their personal leisure time. This would put them in a more comparable position with their husband's or children's leisure time. It would also be possible to lower the age of retirement, that is to say, to organize the "third age" at an earlier stage of life. Another possibility is to recognize the development of the Third World as a most important problem, making it preferable to invest a part of the leisure of post-industrial-society workers in a kind of voluntary work that would be oriented toward the building of an "economy of giving," which will provide the material and human means to mobilize the unemployed time of the underdeveloped societies.

According to Kahn and Wiener, the difference between the average yearly incomes in North America and South America is $2,275. By the end of the century, even after extensive development and aid programs, the difference will be $5,563. This means that the "progress" of affluent society would increase the gap by 100 percent. Would this be a civilization of leisure, or would it rather be a civilization bent on suicide? Is not this dramatic gap as dangerous for post-industrial society as the Chinese having the atomic bomb? We have sought to explain that the passive transformation of leisure time is not the best way to build a more human civilization in post-industrial society. This new society that produces more services and more free time needs to invent an imaginative policy for the utilization of liberated time, to meet the needs of social and cultural development in Third World nations.

Second, what new world view, what new culture will issue from the development of mass leisure in post-industrial societies? What will be the level of cultural values for mass leisure?

There was a time when the progress of culture was more or less identified with the progress of universal reason. According to Marie J. Condorcet, the first duty was "to make reason popular." The spreading of schooling was the consequence of this thought.

In the course of the nineteenth century, there was a reaction in favor of a more concrete culture, closer to technique and to manual work. Philosophers and poets protested against a too-rational, scientific, and technical civilization that ran the risk of forgetting values, values related to the body, passion, myth, spirituality. They juxtaposed culture and civilization.

Finally, cultural relativism made its impact. Culture and nature were contrasted. Each society became more distinct because of the way it managed its environment. The idea of universality was weakening. In industrial society, universal culture was attacked as a class culture.

Today, in the early post-industrial societies, all these questions still exist, but many of the quarrels they aroused are over. Almost all these trends of thought are still alive. Any definition of mass culture which ignores them would remain superficial. But this is not a new problem. What are the new problems of cultural development in the period we are entering? In our opinion, the new problems of cultural development for mass society for the new majority may be formulated by a double alternative in the choice of values. Some of the possibilities are the following: Abstention from former institutional obligations; the invention of new social involvements; the keeping of a balance between the renewed value of professional, familial, religious, and political involvement in the everyday life of the majority of people and the new values of leisure, so that the masses, not only the elite, increasingly participate in the elaboration of their own destiny. On the other hand, the masses could hand over this fascinating but tiresome power to an oligarchy of intellectual specialists, technocrats, or bureaucrats in order to enjoy their increased time of leisure and the consumption of goods, which will be distributed in growing quantities to a growing number of individuals. Of course this danger is still far from threatening the totality of the population of advanced industrial societies. There remain serious inequalities which should be rectified. But this is a problem for a majority, and we must foresee at what stage we shall be in twenty years. If we do not do so today, won't tomorrow be too late? It is well to remember the warnings of American sociologists of mass leisure and mass culture.

FINAL OBSERVATIONS

CHARLES OBERMEYER

CHALLENGES AND CONTRADICTIONS

What is leisure? As is true so often, the Greeks had a word for it. They called it *schole,* and it had several meanings: leisure, or serious discussion, or school. Evidently they believed in schooling people for, or in, leisure, and in serious discussion as one of the best ways of doing this.

Sophocles, as Greek as they came, added a warning: "A purposeless leisure breeds nothing good." Pull all this together and you've got something that begins to make sense out of our problem—which is what you learn to expect from the Greeks if you live with them long enough and get involved with crucial problems. Leisure, they suggest, is not worth having, is even dangerous if it is not shot through with purpose or meaning. Leisure is not something added to life, a busy-idle diversion or a long, drawn-out sigh of relief after work, a sigh that is often indistinguishable from boredom. It is the process that builds meaning and purpose into life, and serious talk, solid exchange of ideas, is one of the best ways of doing this.

Such a process means starting early, in youth. You can't build purpose into a life that has run most of its course without a sense of purpose or meaning except that of adjusting to the survival, security, or success demands of its society. And so leisure, as the Greeks saw, begins in school, and all life is schooling. Youth, through their more sensitive or more disturbed vanguard, are asking today—even demanding—that we help them build purpose into life while they are still "at leisure," so to speak, in school. They tend to wrap the challenge in the word "relevance." Make it relevant; what you are giving us is not relevant or meaningful. And if we counter with a question: How do you know what is relevant, your life is just beginning? they are apt to say, in intent, Maybe, but we are wide open to life, exposed and very vulnerable. We have not yet, most of us, developed the protective coloration and security of the man or woman with a job and family or protection from the draft; nor are all of us scared into acceptance or conformity by the fear of difference or ostracism. Somehow we sense and know that the world we live in doesn't make sense and the way we're going won't be good for our kids. We know enough—you taught us, remember—to know that it isn't necessary and that it could be different. We have a magnificent technology. What's wrong with our sense of human values that we allow it to threaten us with disaster, with pollution and World War III? Why talk about leisure when there probably won't be any for us to enjoy, and surely not for the backward nations who feed their raw materials into our affluence? And what kind of values come out of our affluence anyway?

There you have all the terms of the conference, technology, human values, and leisure, all subsumed in a way under "relevance." Now suppose you ask, How would you make it all relevant, or build meaning into the "schooling" or leisure process of life? That is a tough question and youth cannot be expected to have the answer. All youth knows, often enough, is that what they get does not usually concern them enough to be "grappled to the mind with loops of steel," to paraphrase Shakespeare. Isn't that the responsibility of us older folk who have had more leisure, who are supposed to have weighed things in the balance and thrown away what was found wanting? But we didn't. We collected and collected and put it all in the attic, and divided the attic into more and more subdivisions and departments and separate disciplines. What is more—or worse—we never *really* came together to make sense out of it all, to impose some human form on it so that our youth could build some integrated sense out of it. Obviously, they can't build purpose or meaning into life this way, and equally obviously there cannot be much prospect of a creative leisure; only a dangerous leisure or aimlessness that can fall easy prey to any prowling, power-hungry groups riding roughshod over human values with the ever more powerful and ever more sophisticated machinery of our technology. Once again the three terms of the conference, but in a sinister context. Why sinister? Because contradictions are beginning to show up, and nothing can be more distressing—and challenging.

Let's begin with the contradictions growing out of our technology.

Our technology opens up the possibility of an age of leisure, but it may be destroying our ability to make use of it. It is externalizing the human mind, pulling it into more and more involvement with things, with the shape and price and speed of things, the possession and excitement of things whereas a creative culture— one that will not lead to boredom and ultimate empti- ness—demands an internalization, a building, of the inner life of man. A creative culture demands an in- tensified sense of the importance of the life of the mind, something that is subtly but steadily being under- mined by our machine technology. It demands a deep- ened respect for people, a qualitative sense of man whereas our technology is quantifying our life, chang- ing people into numbers and statistics, into casualties and fatalities.

All of this has been said a thousand times in a thou- sand different contexts. How do we move out of this quagmire of contradictions? How do we set about building the quality that is needed for a creative lei- sure? There is a timeworn answer: By way of educa- tion. But there are contradictions at the heart of edu- cation itself and several of the contributors to the conference were not slow to point them out. Here is Robert Hutchins: "Education has been the processing of the young for the industrial society" and "the multi- versity will have to go with [the industrial state]. . . . At the present time the educational institutions of America are counter-educational." And here are some critical remarks about our education from Robert Theobald: "Disciplines are fragmentations of gestalts and totalities: one can only meaningfully analyze totalities." "[We must] stop ruining people's lives by the education we are giving them." And Joffre Dumazedier of Paris adds this suggestive note: "Social movements of students constitute a revolt against the schooling system especially in advanced industrial society." The conference didn't go as deeply into the contradictions of our educational process perhaps as their provocative criticisms suggested. So let me try to open up a few more approaches to the question of education that remains neutral to the whole problem of leisure and the reconstruction of human values in an age dominated by technology. Creative leisure cannot be modeled on the kind of leisure that intellectuals have enjoyed and developed in the past. Intellectuals could afford to break up the mind into subdivisions and to retire into their separate cubicles, or "wormholes," as they have been called. They had a gentleman's agreement that they would not intrude on each other's baliwick or challenge each other—except over, or rather "between," cocktails or their earlier equivalents.

Students didn't cause any trouble. They were a captive audience carefully regimented by a system of grades, credits, required courses—even the options were carefully structured. In the United States students obligingly called themselves "kids" and the faculty gladly embraced the word.

That was before social problems became issues and began to bite into the conscience of students and to force themselves into the classroom. But social problems were piling up. The dazed and insecure Negroes became proud and insistent Blacks and began pounding on the door. There was much talk about integration, but the colleges had provided no integrated understanding of the roots of segregation. And the contradiction: we were presumably fighting for the self-determination of colored people in Vietnam, while in the United States colored people were fighting us for their self-determination. The problem is intensified by the fact that the United States is "involved" all over the world, and three-quarters of the world is colored, and for our GI's the North and South Vietnamese are "gooks."

It began to become clear that the colleges had no real way of understanding, let alone tackling, the contradictions that were intensifying outside the colleges and impacting the students. The air was full of wars, memories of wars, present facts of wars, and threats of future wars, all mixed into the bitter contradiction that we were preaching peace and desperately *needing* peace while our whole economy and politics were tied into war and preparations for war. In the meantime, our sociology departments could not provide or suggest any adequate texts about war, let alone set up adequate courses. Psychologists took cover, or rather remained under cover. They stuck to their rats and pigeons, their stimuli-and-response techniques, their conditioned reflexes, and their carefully castrated versions of Freud all wrapped in harmless growth and adjustment patterns and techniques—just when "adjustment" was becoming an obscure word for youth.

Now the stench of pollution is filtering through the carefully insulated walls and into the air-conditioned classrooms. The contradiction? That the very technology that provides, advertises, and disseminates air conditioning, detergents, deodorants, and vacuum cleaners is also the technology that pollutes the environment, and the two are in the hands of the same people. And how almost hypochondriacal our industry has made us about purity. But note the subtle contradiction in the phrase "cleaning up."

Our colleges have no departments that can begin to articulate the immense complexity of the problem with its involvement with, and impact on, economics, politics, international relations, psychology, all the sciences, natural and social, ethics—nothing escapes. Just think for a moment of the failure of biology—the science of life, presumably—in the face of the problem. It did not drive home the crucial fact that the environment does not surround the organism but is part of the life of the organism, that life is not lived in an environment but is, in a subtle sense, a creative function of the environment; destroy the one and you destroy the other. Just so, you can't separate the individual from his social environment.

Add the harsh contradiction of poverty, which drives home the unpleasant fact that the gap between advanced nations and backward nations is widening, with the industrially powerful nations, of necessity if not intentionally, draining raw materials from backward nations and forcing them into undifferentiated economic patterns, and the United States itself does not escape the contradiction of poverty. This nation has not solved the problem of poverty, and unemployment is the Achilles heel of our economic system; it is an undesirable and unproductive form of leisure.

All these contradictions—and a dozen more besides —should make it fairly clear that there is as yet no cultural program that could put a cultural "body" into leisure, or that can even help us define what leisure really is or could be. They make clear also that our colleges are not yet ready to do the job. All we know is that the development of leisure will take place in a world where very few people, relatively speaking, will have it, for the whole United States of America makes up only 6 percent of the world's population. If the first segment of our population to move into this increased free time is mostly workers, as Arthur Schlesinger tells us, they can be left to enjoy it as they see fit. What they do or do not do with this new free time is, in part, their own concern and, in part, of little moment in the great ocean of human tensions and contradictions whose roar disturbs our peace even when we seem or pretend not to hear it. This may seem like an avoidance or dismissal of the issue. Not so, if we transfer the idea of leisure, as we have suggested earlier, to a plane of the creation of the kind of values and understandings that will give us a renewed grip on life just when we seem to be losing it—especially among the intellectuals. Those who always have had some leisure of their own are faced with the challenge directly, and they had better face it and not crawl back into their academic worm holes. For the world of youth, which has provided most of them with a function *and* a living in the past, is changing under their feet. The war cry of youth has become relevance, and there can be no valid objection to this demand for relevance, especially since so many teachers in colleges have lost grip on their own subject matter. They are not certain whether their subject matter has solid contemporary value for the students they are teaching or why they themselves are teaching it—except that it is a prolongation of their own Ph.D. involvement, part of the process of self-perpetuating Ph.D.'s. To grapple with this issue of relevance in the curriculum is to go to the heart of the problem of leisure. Can we rescue leisure from the nonmanipulative activities to which it has been relegated in the United States and move it to a new creative area of activity and social value away from the private area of speculative thought and contemplation which some of the conferees suggested as its logical domain? Doesn't a musician "manipulate" an instrument, a poet words and images, a painter, paints and brushes, a sculptor, stone, clay, wood, or metal, a philosopher, thoughts or ideas? And what content would there be to leisure without such hard-working manipulators?

Now in the past, speculative thought and freedom for contemplation were a privilege of the few who were exempt from hard labor. This had a double drawback. On the one hand, the thought and art and literature these privileged few were involved with were products of the hard work of others, and grew out of the artist's or thinker's intense involvement with his time and society. On the other hand, the enjoyment of the artist's or thinker's work tended to take place in the privacy of a study, remote from the stir and bustle of the marketplace.

But there is some reason for the distrust of manipulative activities among the more sensitive and thoughtful members of our world. Ours is essentially a utilitarian world of means rather than of human ends, a world in which it becomes increasingly difficult to think of and to treat people as ends in themselves, however much they may need to be treated as such. Ours is a world in which we all feel manipulated and transfer that treatment to others; a world in which children tend to become a means for the realization of the dreams of a parent or a compensation for his failure, with love and recognition dependent on success. At the other end of the scale, it is a world where old folk can no longer be considered as a means toward anything useful. They can no longer be manipulated, and here manipulation gets confused with "service" and the old no longer serve—the opposite of the situation in primitive societies where the old are indispensable depositories of wisdom and experience.

But a contradiction is eating into the situation today. This sense of uselessness is catching up with the young people and is the projected lot of millions in an anticipated age of leisure. So what lies ahead? What will become of individual and social relations in the new society? And what will happen to religion and spiritual activities which seem to be declining? These are some of the questions raised by Joffre Dumazedier of Paris. He has supplied an answer to his own questions: "The weekend activity of young people [in France] tends to become a kind of artistic and intellectual milieu, a time for brightening life with exceptional communications between individuals and groups." This approach to the problem of what to do with or about leisure seems to me pregnant with possibilities, not only because it grows spontaneously out of the teeming mind of youth, but because it corresponds to what is already happening to our own American society. It is being transformed into a society in which more and more efficient machines produce the "goods," and more and more men and women provide the "services" and develop new services. It is along this way, I think, that the future lies.

This does not imply service in the pollyanna, guilt-ridden sense, where business pretends to "serve" the community, or churches or politicians preach the need for service. It is service as an activity that grows freely out of the community, an activity that should grow logically out of its leisure. Youth gets involved with it all over the world; this service is a matter of group activity. Mental health, as we are learning to realize more and more, is coming to be a function of group therapy.

With the concept of "group" we see prospects of transcending collectivism or mass-mindedness, which is so distressing a symptom of our industrial age and which has been so tragically exploited by the paranoid power cliques of recent history. Dumazedier himself is worried about it and borrows Riesman's other-directedness to describe what he sees as an inevitable product of television, movies, games, etc. But the industrial society that we are supposed to be outgrowing was not only "other" directed, it was completely "other" controlled. The new life-style that youth and the blacks are developing in their struggle is at all events different and, I think, pregnant with new and exciting possibilities for individual and social relations.

In this context I would like to turn to Robert Theobald's claim that each individual is unique and needs diverse communities to develop his uniqueness. If we are unique, the average person is not encouraged to develop or express his uniqueness. He is trapped in a collectivity that paralyzes him and frustrates the social self, the democratic self, the need of the "person" in him to belong to an organized grouping of people that gives him freedom because he belongs, and functions together with the freedom to move in and out of the group at will.

There are many possible groupings that suggest themselves. Some are actually springing up in a hundred different contexts all over the world. Let me just draw attention to one of them. It has a magnificent if sporadic past, a very active present, and what seems like a profoundly significant future—if the world we live in will allow us to continue to struggle for the world we want to live in, the world of creative leisure. I am thinking of the theater. The theater has always involved democratic group effort; it has given a group-imposed, disciplined freedom to the actor, and a self-imposed but group-controlled freedom to the playwright. It has always integrated many different skills and has always involved—and depended on—the larger community, the audience. Anyone who has followed the recent developments of the theater in the world at large and on the hundreds of theater-conscious campuses of the United States, not to mention the mushrooming of theater groups in our great cities and the growth of vitality in the movies, will sense that something very exciting is getting underway. And all this is happening before the so-called age of leisure has gotten started, and in spite of the obstacles to real dialogue that all establishments—nations, churches, colleges, corporations—have tried to roll in the way, despite their continued statements to the contrary.

Theobald sees no signs of "crash courses in reality." But why look to courses in colleges? The reality *outside* the colleges is crashing right through the courses, and disturbing fact is in the saddle. Robert Hutchins longs for the time when "computers and other devices could make every home a learning society." Fine! But who controls the computers? If cybernetics is an extension of the human mind into machines, the reverse is also probably true. How will we give the mind back its autonomy and dignity? That is where the spontaneous group activity comes in with the implication of a new "group theater" of the mind. Here you are out of the reach of computer control, *and* you are out of the occlusive and exclusive protection of the family. The other-directedness becomes democratically accepted and sought-after group direction, with a release of emotions and imagination and a control of them that can push the neurotic, alienated individual on the way to becoming a "person," conceivably, I suggest, the finest product of leisure.

This concept of a theater of the mind, of dramatized thinking and feeling in public, of involving the public in the process, can be extended in a hundred different directions. It can mean moving into colleges from their surrounding communities, and moving from students into faculties, and moving across national borders on the wings of a technology we have already developed but have lacked the courage and imagination to exploit for essentially human purposes. We stop here. We are getting close to what some readers may call a romantic picture of our possible future. Let me substitute the word "ideal" for romantic and suggest that "ideal" may be just what youth wants, a picture or image of purpose and meaning to be built into life by cooperative effort, which will act as a spur from behind at the same time that it is a powerful pull from what lies ahead, from what youth realizes—as yet somewhat vaguely—as a possibility *and* a necessity. Youth sees the ideal as an imperative that says "you *must*, therefore you can"; not, as our technolgical power is seductively whispering in our ears, "You can, therefore you must—or why not?" We could add that the great dramatists of the world seem to agree that to frustrate the dreams and ideals of youth—the romance, if you will—is to move into the dark areas of the mind where the Iphigenias and Antigones, the Romeos and Juliets, the Desdemonas and Cordelias, the millions upon millions of would-be "patriotic" young men have been sacrificed on the altars dedicated to the ugly, frustrated power wills of their elders.

Now if theater is springing up in and outside colleges, is developing a momentum of its own, is extracurricular in more senses than one, what about the other parts of our cherished curriculum, the pride of our colleges if not necessarily of our culture? If they are not reevaluated, transformed, made relevant, how can they function in the building of potent new human values, without which leisure is an empty word, and without which the erosion of our culture by the depersonalizing collectivism of our technology goes on apace? There is a dangerous, *positive* feedback in our technological world, and it tends to gather momentum without stop signs or checkpoints.

This brings us to the heart of the challenge. Those anticipating or enjoying leisure will have to be shaken out of their hope of leisure as

A bower quiet for us, and a sleep
Full of sweet dreams, and health, and quiet breathing.[1]

Leisure is decidedly not an area for the nonmanipulative activities of outsiders, aliens. It will have to become an arena where we do battle with the value-distorting powers. There will be no meaning to leisure, *and therefore to life,* until we slay the dragons of antilife, the dehumanized commercialism and impoverished collectivism and the final negator of the human spirit, the manmade threat of total destruction. That implies for colleges a reassertion of the central role of the human

---

[1] Keats, *Endymion.*

mind, of its philosophic energy to integrate and unify its own content around crucial human values. That is a terrific task today when that content is so apparently overpoweringly varied and distressingly and dangerously disconnected.

It implies a reassertion and dissemination of the primacy of ethics for this philosophic energy—the central, motivating drive in Plato and Spinoza and Kant, and frustrated in Hegel, as the modern Nation-State and the Industrial Revolution began to take over. And the building and assertion of ethics in the life of nations is a task that will demand all the spare energy we can muster. It implies an imaginative broadening and emotional deepening of the historic understanding, a sense of context and relatedness not only of interaction but of essential relatedness so that the tragic failures of history—almost the whole of history—may be seen as failures or breaks in relatedness, something that pollution and war should make clear enough. It implies a vitalizing of art as a forming function of the emotional and imaginative life of the community, a tying of the intensive style of the individual artist into the extensive life of the people around him. This is what all great art has always been in intent, but only rarely, except in a primitive community, has it become the proud articulation—in all the senses of that word—of a society. The reason may be that a society finds it very difficult to become a community; it is torn apart by the anti-democratic power urges that spring up within it, and which, as the society becomes complex, "can get away with murder," go unchecked, often unnoticed, and if noticed, emulated. Here is a suggestion for the future role of colleges: to build creative communities within the nation and move the energy that this effort releases into the larger society. It can, as we have suggested, begin in the theater. Great theater proves convincingly that great art is the product of a moment of heightened solidarity in the life of a community, however small. As small as a college, if you will. It implies a new synthesis of the sciences, a reconstruction of the concept of nature, with all the sciences converging as separate disciplines but inseparable dimensions in the understanding of nature. We must drive home the basic fact that the human mind is the creator of the sciences and cannot, therefore, be analyzed by or contained within the framework or categories of the sciences. This, it seems to me, is of paramount importance today where our sciences subserve our technology and our technology overwhelms and distracts our minds. In our practical American world where we are subtly being manipulated by the managers of the techniques of manipulation, with our science turned into technology and our technology serving the sinister purposes of power, this will be no easy or leisurely operation.

This question of power was raised by Robert Hutchins when he asked: "What do we do with power?" He didn't attempt an answer. The answer lies, I suggest, in the sensing and deepening and widening of the democratic spirit, which is the only antidote to the abuse of power. It is the continuing struggle for involvement and participation as technology enlarges the available "cake" and opens up new potentials. This struggle is going on all over the world. In the United States youth does not seem to be too confident that this historic land of opportunity is really extending opportunity as widely as it could and should. What is more, they are keenly aware, many of them, of the corruption that attends on power. They sense that the struggle is ultimately for the "soul" of man, a struggle to give man back a soul, to build new meaning into life when life is running out or being stolen from them by the power-drunk masters of the world and its technology.

This then is the leisure concept as it is shaping up: the building back of purpose into life rather than the mere enjoying of life. It is the purgatory of struggle that lies close to hell and a long way from heaven; and a long way, too, from the quiet, contemplative life of scholars and the many harmless hobbies of the mind. The mind is being forced back onto the main road of relevance; crucial issues have become *crucial* issues. To end where we began, the Greeks knew this. Leisure, for all their glorification of it, was built on slavery and torn to shreds by suicidal civil wars and cruel contradictions. Between them and us lies their own intensely tragic sense of life—tragic, not defeatist; and it is the courage and wisdom of tragedy that we lack in our culture. Without that courage and wisdom our leisure is an empty dream, a "purposeless leisure" which breeds nothing good. Something of this is breaking through, not surprisingly, in many of the exciting movies of today.

The Eastern sky is red; it may not be with blood.

APPENDIX A

THE ALLOCATION OF LEISURE TO RETIREMENT [1]

Juanita M. Kreps

For purposes of analyzing . . . issues it is necessary to make some projections of future leisure trends, both in total amount and in the possible range of forms it may take. Assumptions must therefore be made as to the growth in productivity per man-hour, labor force size, etc. In table 1, the 1965 projections of gross national product (GNP) made by the National Planning Association (NPA) are used, the basic assumptions being,: between 1965 and 1985 the growth rate will be 4.1 to 4.2 percent per year; population will grow

[1] From "The Leisure Component of Economic Growth," by J. M. Kreps and J. J. Spengler, in *The Employment Impact of Technological Change*, Appendix, vol. II, National Commission on Technology, Automation, and Economic Progress (Washington, D.C.: Government Printing Office, 1966), pp. 353-97.

TABLE 1.–Prospective growth in productivity and possible uses of released time

Year	Possible increases in real GNP (1960 dollars)		Alternative uses of potential nonworking time				Education and training	
	GNP (billions)	Per capita GNP	Total number of years	Retirement age	Length of workweek (hours)	Vacation time (weeks)	Labor force retrained[1] (percent)	Years of extended education
1965	$627.3	$3,181	. . . . . .	65 or over	40	3	. . . .	. . . .
1966	655.6	3,280	2,245,542	65	39	4	2.9	1.2
1967	685.6	3,382	4,655,526	63	38	7	5.0	2.4
1968	707.1	3,290	6,910,648	61	36	7	8.7	3.4
1969	745.3	3,573	8,880,092	59	36	8	11.1	4.2
1970	779.3	3,690	11,263,301	57	34	10	13.8	5.1
1975	973.4	4,307	23,135,642	50	30	16	26.2	9.4
1980	1,250.2	5,059	35,586,729	44	25	21	37.2	13.8
1985	1,544.5	5,802	47,200,158	38	22	25	45.2	17.5

[1] Figures are in addition to the number of workers now trained in public and private programs.
Source: GNP projections and employment data from National Plan- ning Association, Report No. 65-1, March 1965. Labor force data for other computations taken from *Manpower Report of the President,* March 1965, p.248, table E-2.

by 1.5 percent annually; unemployment will average 4.5 percent.

In order to show potential GNP on the assumption of no change in working time—NPA estimates GNP on the basis of a decline in working time of one-half of 1 percent per year—the GNP figures used here, corrected for this decline, are slightly higher than the one derived by the Association. Assuming no change in working time, the GNP at projected rates of growth would approximate $1,544,500,-000,000 in 1985, about 2⅓ times it present level in 1960 dollars. Per capita GNP would rise from $3,181 to $5,802, or more than 80 percent, despite the increased population size. Less rapid increases in aggregate and per capita GNP than these projections indicate may occur, of course, particularly if shifts in labor force composition (from manufacturing to services) are sufficiently rapid to slow the overall rate of productivity growth.

These increases in total and per capita GNP are possible, then, if working time of roughly 40 hours per week for an average of 49 weeks per year is continued. At the other extreme, if one supposes that all growth except that amount necessary to hold per capita GNP constant at $3,181 is taken in leisure time, the possible increases in free time are indicated in the remaining columns. The workweek could fall to 22 hours by 1985; or it would be necessary to work only 27 weeks of the year; or retirement age could be lowered to 38 years. If the choice were made to divert the new leisure into retraining, almost half the labor force could be kept in training; if formal education were preferred, the amount of time available for this purpose might well exceed the normal capacity to absorb education.

It is, of course, not likely that the workweek will drop to 22 hours or that retirement age will decline to 38 years. Nor is it probable that during the next two decades workers will continue on their present schedules, thereby taking all productivity gains in the form of a greater quantity of goods and services. If, instead, two-thirds of the output growth accrued as goods and services and one-third as leisure, GNP would rise to more than a trillion dollars by 1980, and to $1.3 trillion by 1985. Per capita GNP would

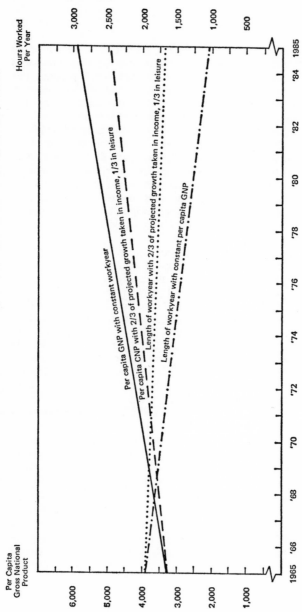

FIGURE 1—Alternative Uses of Economic Growth Per Capita Gross National Product and Hours Worked, 1965-85

Per Capita Gross National Product

Hours Worked Per Year

Per capita GNP with constant workyear

Per capita CNP with 2/3 of projected growth taken in income, 1/3 in leisure

Length of workyear with 2/3 of projected growth taken in income, 1/3 in leisure

Length of workyear with constant per capita GNP

Source: GNP projections and employment data from National Planning Association, Report No. 65-1, March 1965. Labor force data for other computations taken from Manpower Report of the President, March 1965, p. 248, table E-2.

increase to more than $4,400 by 1980 and to approximately $5,000 in 1985. (See fig. 1.)

The leisure which accounts for the remaining one-third of the growth potential could be distributed in any way or in a combination of several ways. Different priorities would be assigned by different persons. If it is conceded that present unemployment is due in some significant degree to qualitative deficiencies in the labor force, the first priority might be assigned to job retraining. Hence, a policy dicision could be made to retrain a minimum of 1 percent of the labor force annually, taking the necessary time from that freed or released by the growth in productivity. A second order of preference might be an increase in vacation time, at least until an average of 1 additional week accrued to the worker. By 1968, these two goals—retraining 1 percent of the labor force and increasing vacation time by 1 full week—could be attained. If, after these achievements, some leisure gains were taken in the form of reductions in the workweek, working time per week could start by declining about one-half hour in 1969, the decline increasing to 2½ hours by 1980. (See table 1.)

Alternative allocations of leisure in the period 1980-85 might be as follows: Given a $4,413 per capita GNP in 1980, achieved with a 37½-hour workweek, a 48-week work year, and providing retraining for 1 percent of the labor force, society could choose to retrain much more heavily (4.25 percent of the labor force per year) or, alternatively, could add 1½ weeks per year in vacation. In 1985, when per capita GNP should reach about $5,000, the choice could be between retraining almost 7 percent of the labor force annually or taking an additional 3 weeks of vacation. Obviously, other choices could be made, involving a further reduction in the workweek, a lowering of retirement age, or an increased educational span for those entering the labor force.

The relevant considerations are at least threefold. One, the total amount of free time made available by the anticipated improvements in output per man-hour is extremely great, even when allowance is made for quite rapid rises in real GNP or even in per capita real GNP. Two, the allocation of this leisure is in itself quite important, given the different degrees of utility man may associate wth different forms of leisure. Three, the distribution of leisure, being quite unevenly spread over the entire population, requires further consideration. For although the unequal distribution of income among persons has received great attention, it might well be true that that portion of economic growth accruing to man in the form of leisure has in fact been apportioned much less evenly than has income. Questions relating to the total volume, the forms, and the distribution of leisure are of some significance in estimating future potentials for growth in output, and particularly in determining the composition of that output.

APPENDIX B

FREE CITY
FESTIVAL **OCT. 30 & 31**

8 BANDS

USF

FOOTBALL FIELD

Free

Yippie!

FREE CITY MUSIC PRESENTS...
CELEBRATION OF LIFE

Free, Free, Free! CELEBRATION OF LIFE sponsored by Free City Music will begin Friday night, Oct. 30 after a rally called by S.M.C. The festivities are free & open to all. Eight bands, as of Monday, have agreed to play free for us and the list is growing. We invite all bands and musicians to come. We need volunteers beginning Friday morning to help set up the stages, lighting, power, toilet facilities, OD tent, and Free Kitchen. Please bring tents and sleeping gear and whatever else you need. Free, Free, Free! Keep on Trucking!

APPENDIX C

CENTER FOR STUDIES OF LEISURE

The Center for Studies of Leisure is an integral part of the University of South Florida in Tampa. It is perhaps the only university agency devoted entirely to the subject of leisure in the broadest sense: a concern with the total pattern of work and nonwork trends of the post-industrial society related to cybernation, increases in bulk time, flexible work patterns, urbanization, changing values, public policy, expenditures for recreation, and new demands on education and other social institutions. This is done through conferences, consultations, field research, lectures, writings and newsletters, workshops and seminars.

Communications are underway with many nations to create an international collection of films on leisure; a quarterly *Newsletter* is widely distributed; an innovative Master's program in the study of leisure is being prepared. The Center represents the United States in a nine-member team of scholars, including France, Canada, West Germany,

Sweden, Switzerland, Poland, Bulgaria, and Czechoslovakia, designated as a research committee of the International Sociological Society and UNESCO.

Three advisory groups work with the Center: a University group of faculty and students, a community-regional group of civic leaders, and an international board, which consists of Joffre Dumazedier of the Centre d'Etudes Sociologiques, Paris; Paul Lazarsfeld, Columbia University; David Riesman, Harvard University; Alphonse Silberman, University of Cologne; and Alexander Szalai, Deputy Chief for UNITAR, United Nations.

INDEX